U.S. MILITARY GROUP PHOTOS

Eric "Superman" Sturgess

Rose Of Sharon
PUBLISHING

ERIC "SUPERMAN" STURGESS

Printed Worldwide
First Printing 2025
First Edition 2025

10 9 8 7 6 5 4 3 2 1

Rose of Sharon Publishing books may be ordered through booksellers or by contacting:

Rose Of Sharon Publishing
roseofsharonpublishing.com
P.O. Box 186
Floresville, Texas 78114

ISBN: 978-1-970335-12-5 Paperback

Preface

I served four years in the United States Coast Guard and was honorably discharged upon completion of my enlistment. My boot camp photo was lost in a move over 20 years ago, and that really bothered me. I tried for many years to find a copy of it online, but all of my searching was in vain. A few months ago, I came across a copy of it, and it was a great day.

These pictures of military boot camp classes and technical schools hold a great deal of significance for those who were photographed in them. The only way to obtain one of these original pictures is after the owner has passed away. There is no other way they would ever part with one of these significant moments in their lives when they were young.

These military class photos are a snapshot of time that captures these men doing their best to honor their country and serve their fellow Americans. Obviously, depending on the date of the class photo, most of these men are no longer with us here on earth. These photos and the people in them deserve to be seen by others and their lives not forgotten in the dustbins of history. It just reminds us how fleeting our lives are. Nobody is guaranteed tomorrow, and our youth quickly fades to old age if we are blessed to live that long.

God bless you
Eric Sturgess
Romans 10:9-13

Table of Contents

Preface.....3

Introduction.....5

Photographs.....7

About the Author.....34

Other Books.....37

Introduction

These photographs are more than just images—they are living records of courage, discipline, and sacrifice. Each one captures a moment when young men and women stood together in uniform, united by a single purpose: to serve their country. The faces in these group photos reflect a range of emotions, including pride, uncertainty, determination, and hope. They are the faces of America's defenders from years gone by.

Every photograph tells a story. Behind each face is a life—someone's son or daughter, someone's friend, someone who chose to dedicate a part of their youth to something greater than themselves. Some came from small towns and farms, others from bustling cities and coastlines. They represented every corner of the nation, every background, and every dream. Yet, for a moment in time, they stood side by side—equal in purpose, bound by service, and ready to answer the call.

The settings may vary—Army barracks, Navy decks, Air Force hangars, Marine Corps fields, Coast Guard stations—but the sentiment is the same. These photos mark the completion of training, the transformation from civilians into soldiers, sailors, airmen, marines, and guardians. They capture that powerful instant when individuals become part of a team—a unit forged by shared effort, endurance, and belief in something bigger than self.

For those who appear in these images, these class photos were more than souvenirs. They were symbols of achievement, pride, and a sense of belonging. Framed in homes, kept safe in albums, or tucked away in drawers, these pictures carried deep meaning. They reminded each veteran of who they once were—their youth, energy, camaraderie, and the purpose that guided their steps.

Over the decades, many of the faces in these photographs have grown older, and many are no longer with us. Yet their presence endures through these preserved moments.

Time may fade ink and paper, but it cannot erase the spirit of those who served. These images remind us that the strength of a nation is measured not just by its victories, but by the men and women who stood ready to protect it.

This collection brings those memories back into the light. Each photograph is a window into history—a quiet testament to dedication and unity. Together, they form a portrait of a generation that believed in service, honor, and country.

As you turn these pages, look closely into their eyes. You will see courage. You will see youth and conviction. You will see ordinary Americans who took an oath and fulfilled it. They may not have known what awaited them, but they stood together nonetheless.

This book is dedicated to them—the men and women of the United States Armed Forces—who trained, served, and sacrificed so that others could live in freedom. Their faces tell a story that words alone could never capture. And through these photographs, that story will never be forgotten.

Photographs

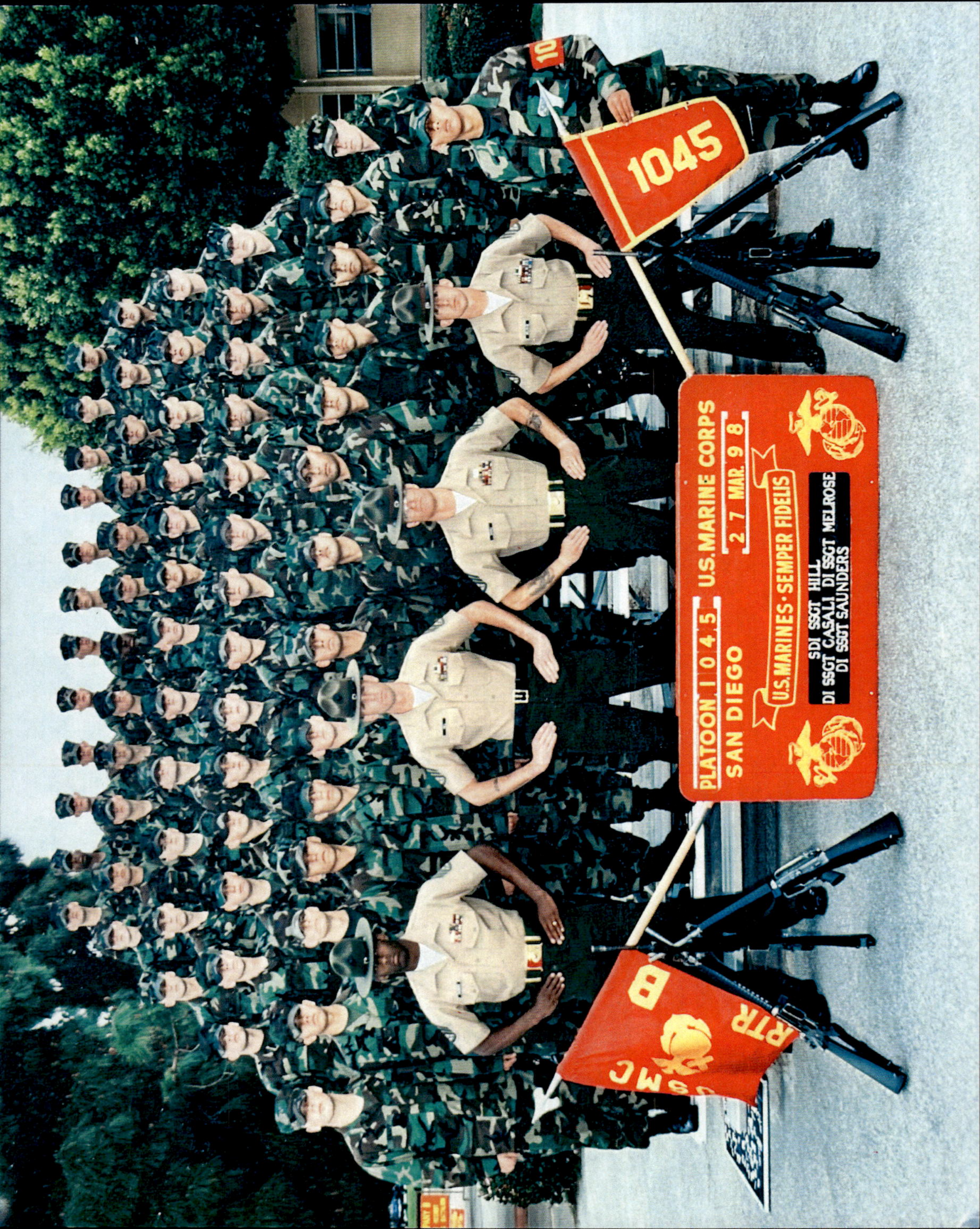
1045
PLATOON 1 0 4 5
U.S.MARINE CORPS
SAN DIEGO
2 7 MAR. 9 8
U.S.MARINES·SEMPER FIDELIS
SDI SSGT HILL
DI SSGT CASALI DI SSGT MELROSE
DI SSGT SAUNDERS
B
RTR

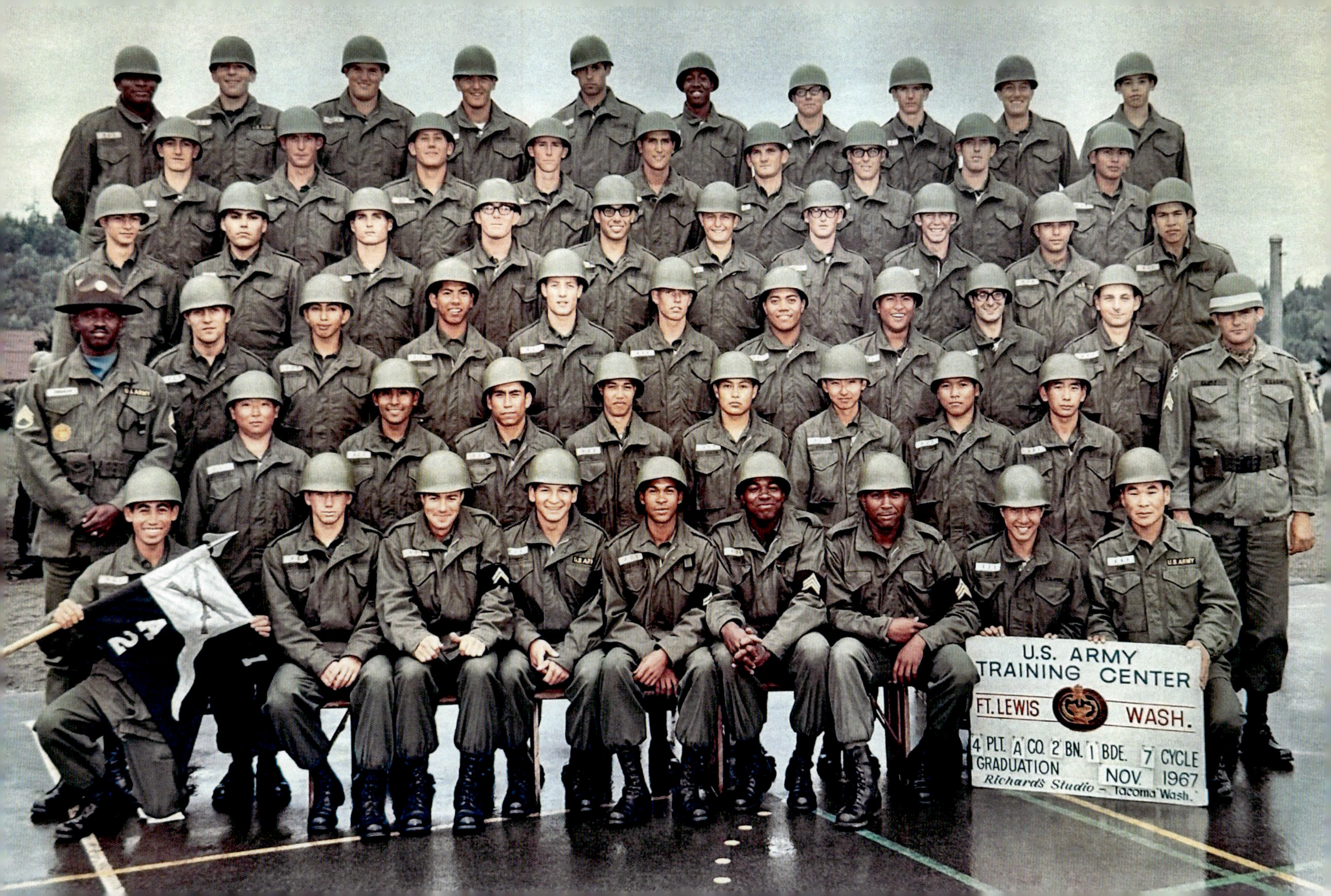
U.S. ARMY
TRAINING CENTER
FT. LEWIS
WASH.
4 PLT. A CO. 2 BN. 1 BDE. 7 CYCLE
GRADUATION
NOV. 1967
Richard's Studio - Tacoma Wash.

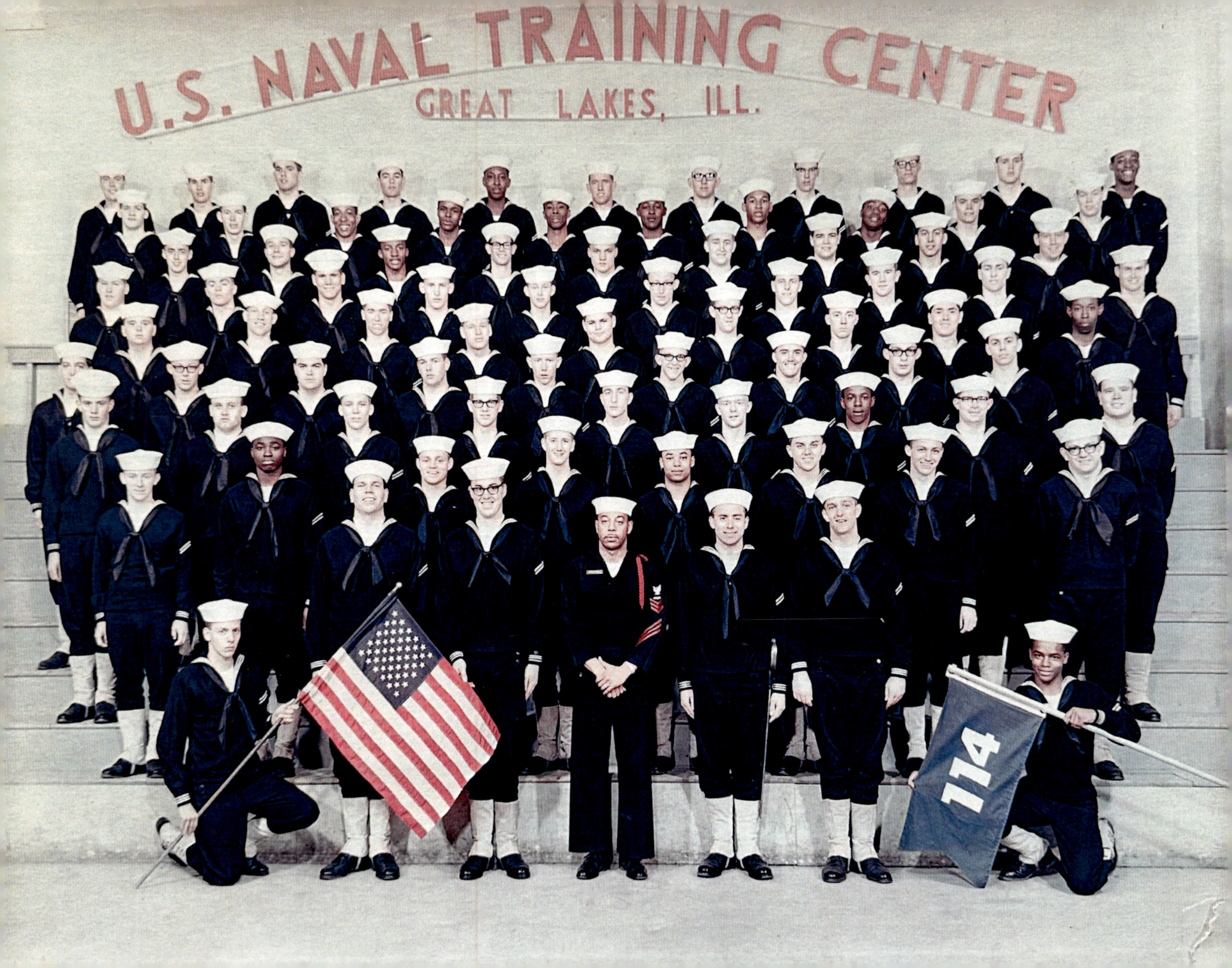
U.S. NAVAL TRAINING CENTER
GREAT LAKES, ILL.
114

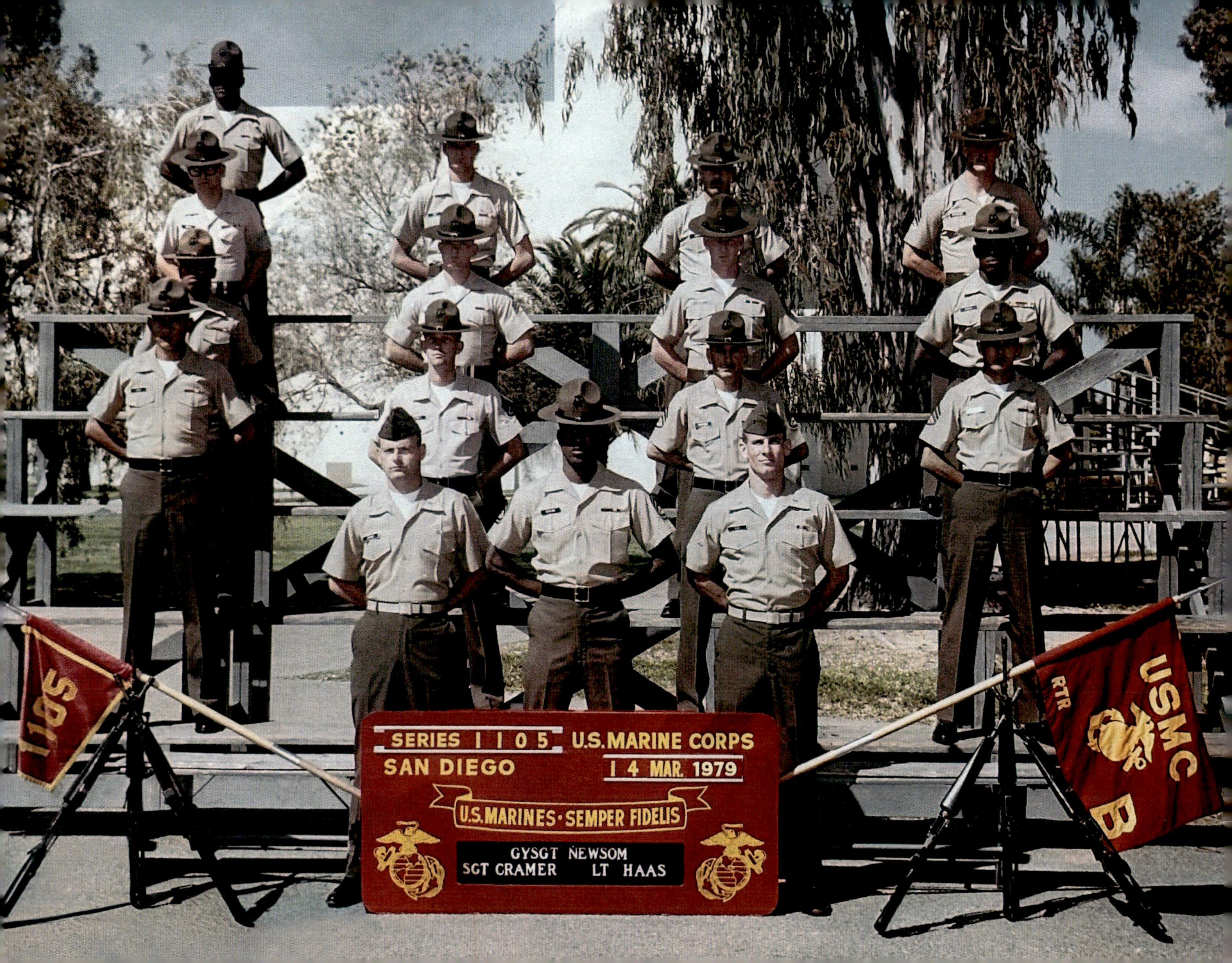
SERIES 1105
U.S. MARINE CORPS
SAN DIEGO
14 MAR. 1979
U.S. MARINES · SEMPER FIDELIS
GYSGT NEWSOM
SGT CRAMER
LT HAAS
USMC
RTR
B

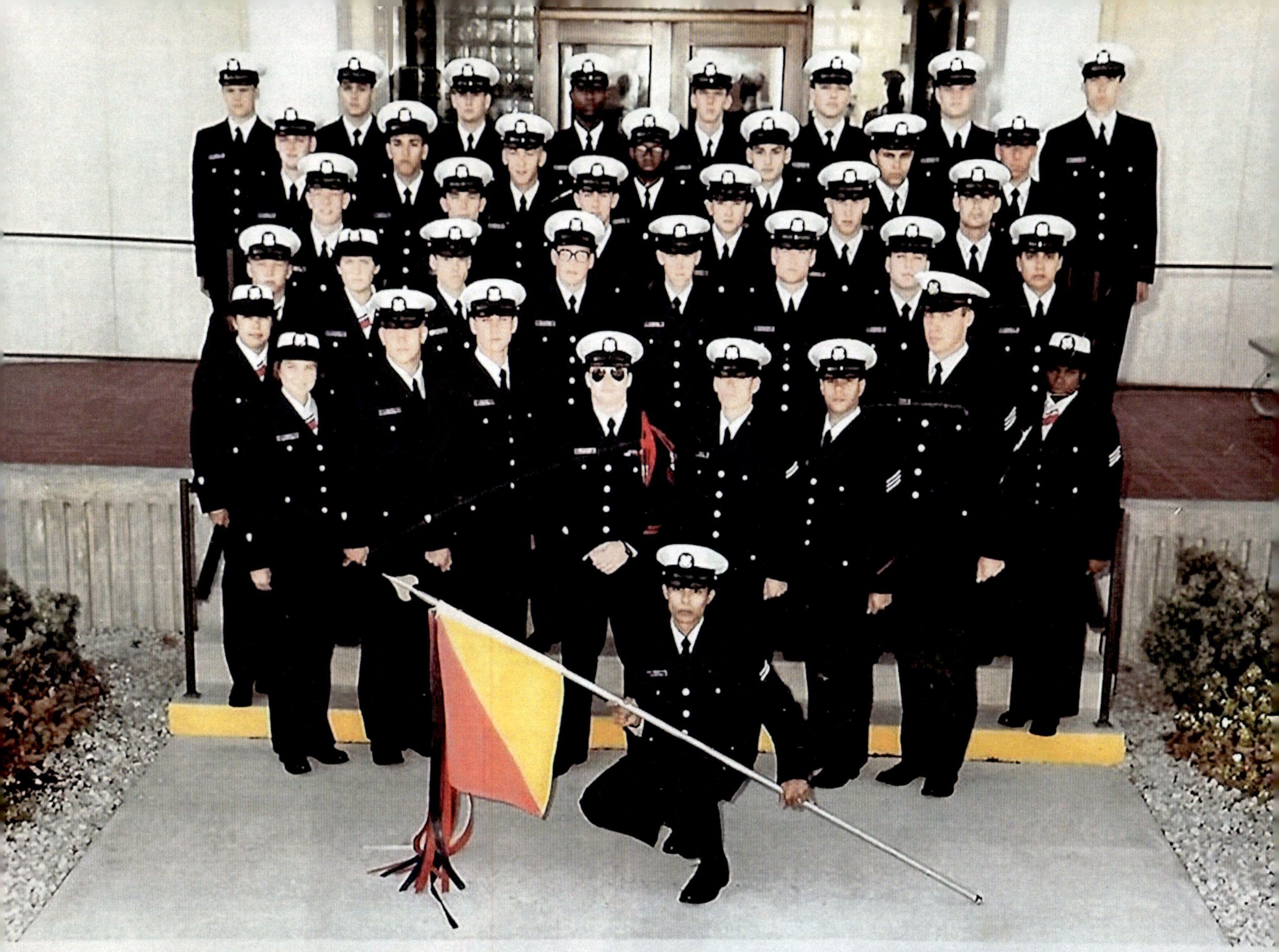

U. S. COAST GUARD TRAINING CENTER CAPE MAY N. J.

COMPANY O-134 RM1 WILFORD 16 NOVEMBER 1990

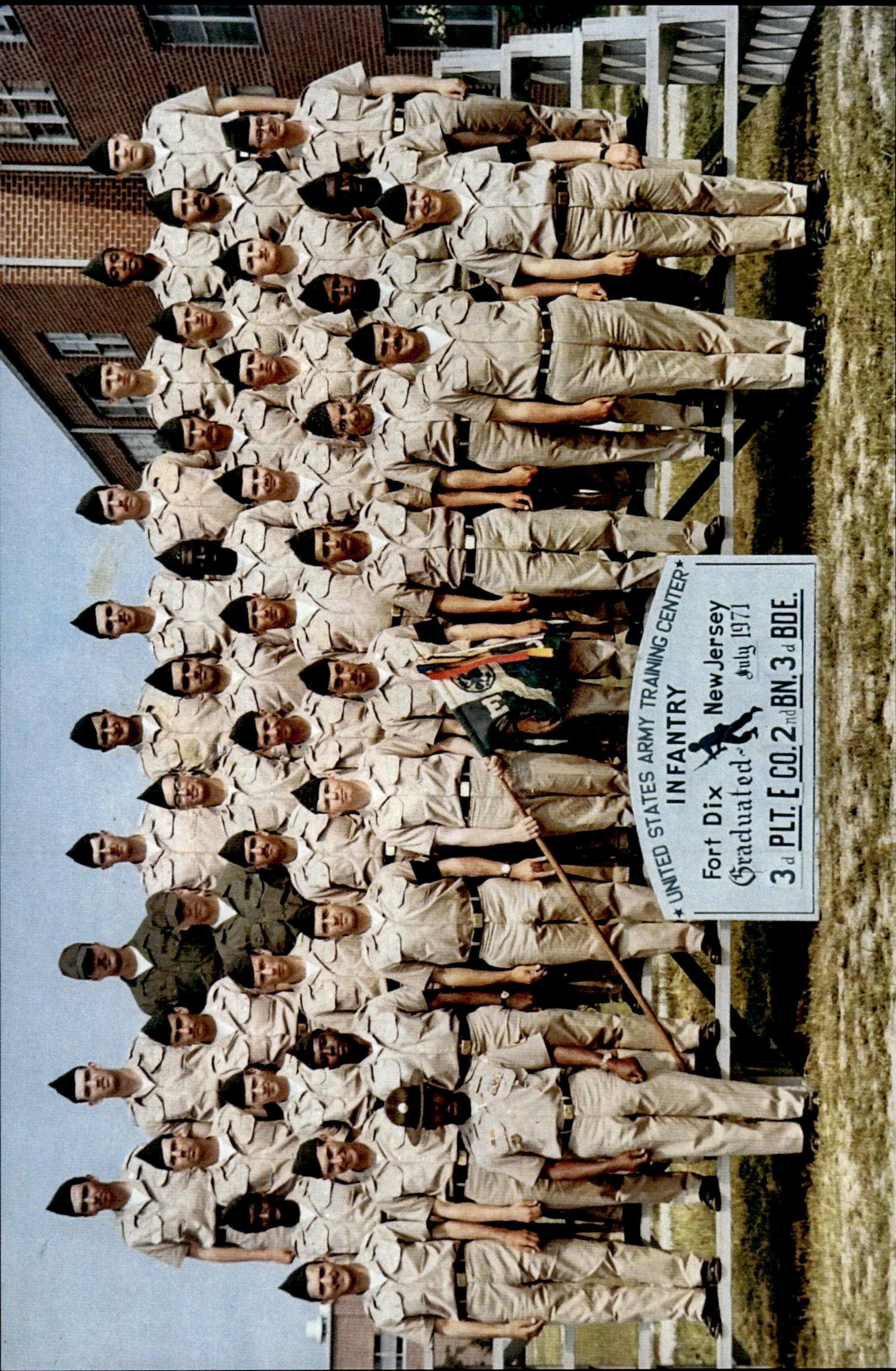
UNITED STATES ARMY TRAINING CENTER
INFANTRY
Fort Dix New Jersey
Graduated July 1971
3d PLT. E CO. 2nd BN. 3d BDE.

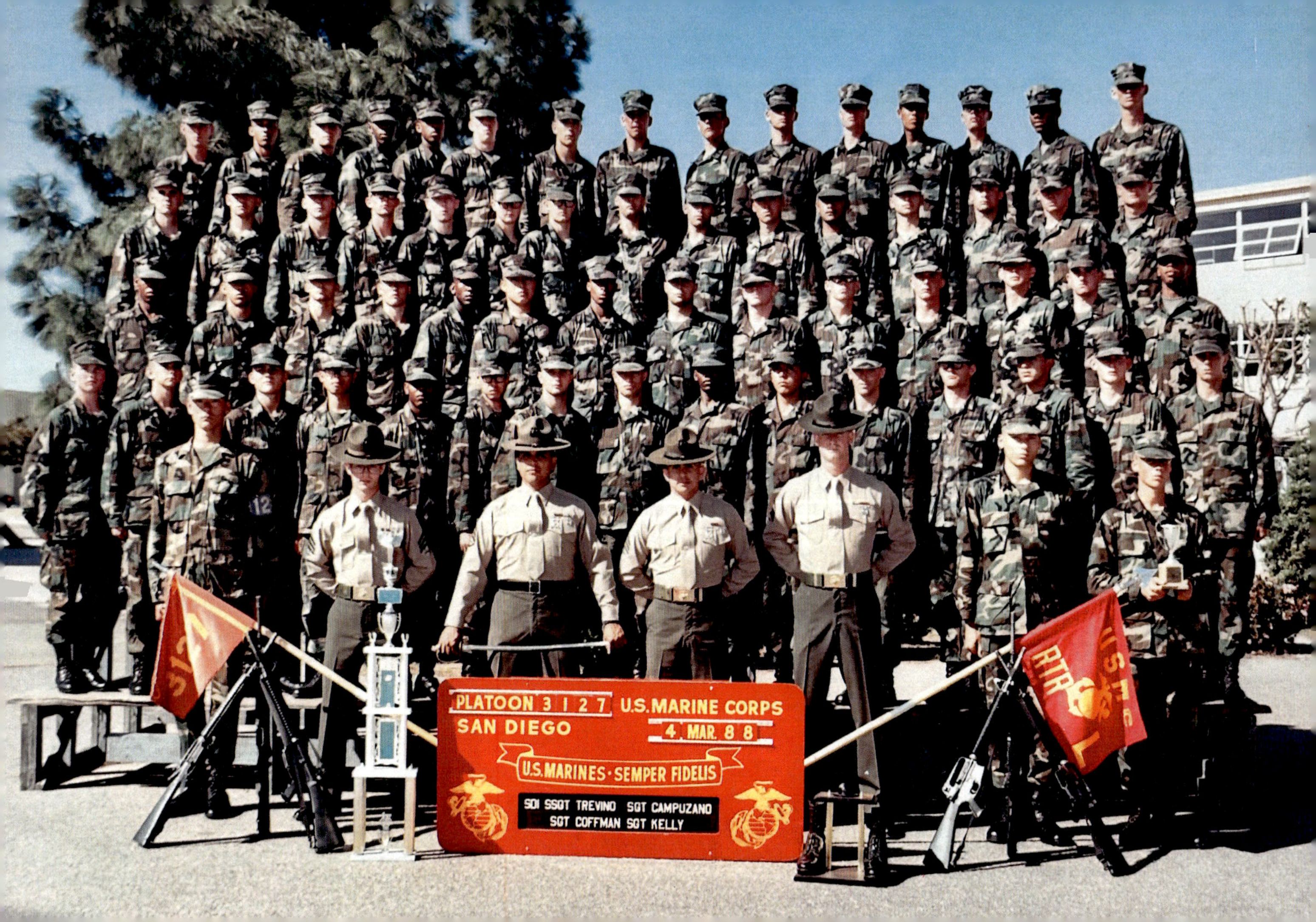
PLATOON 3127
U.S. MARINE CORPS
SAN DIEGO
4 MAR. 88
U.S. MARINES · SEMPER FIDELIS
SDI SSGT TREVINO SGT CAMPUZANO
SGT COFFMAN SGT KELLY

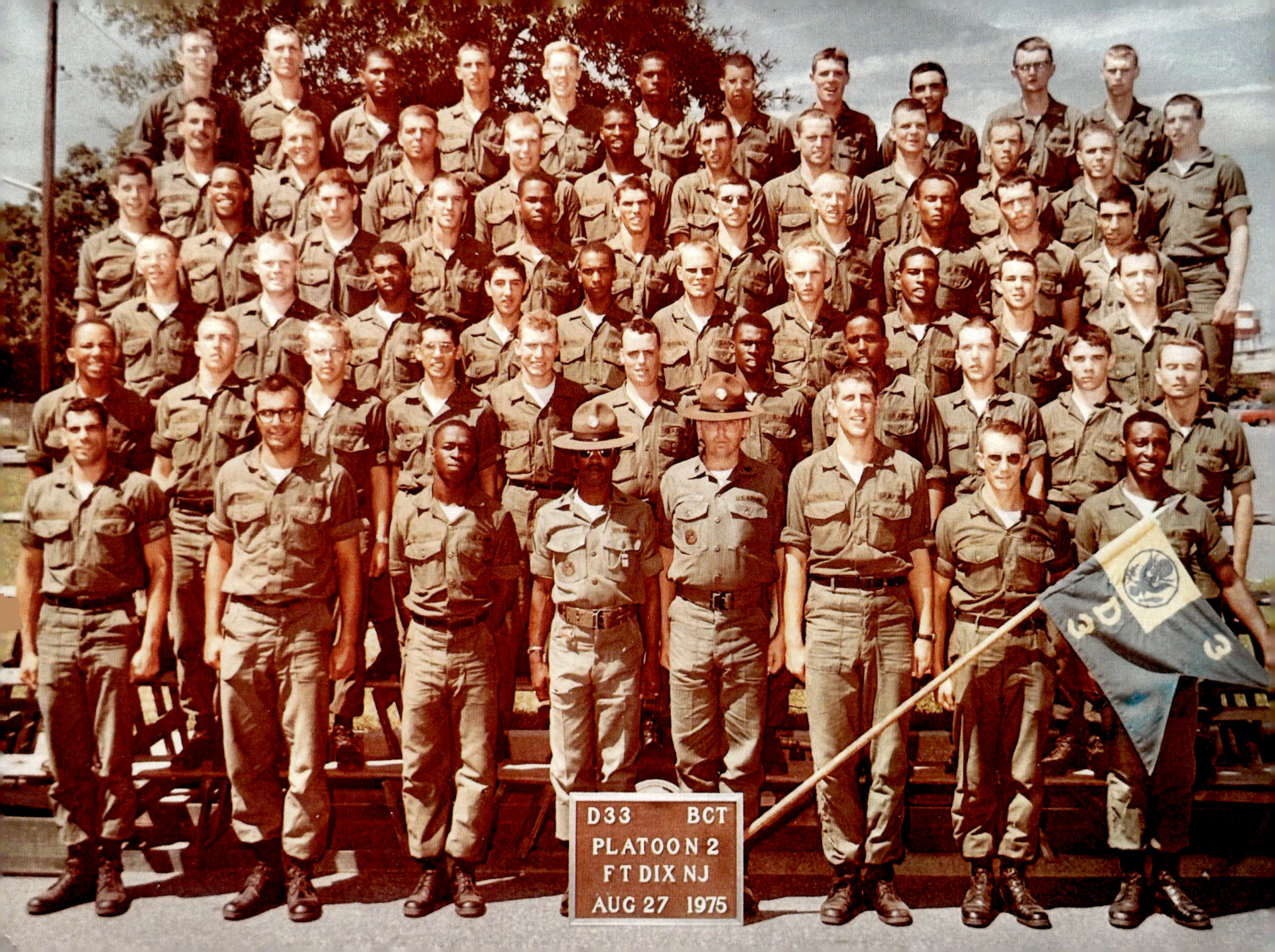
D33 BCT
PLATOON 2
FT DIX NJ
AUG 27 1975
D
3
3

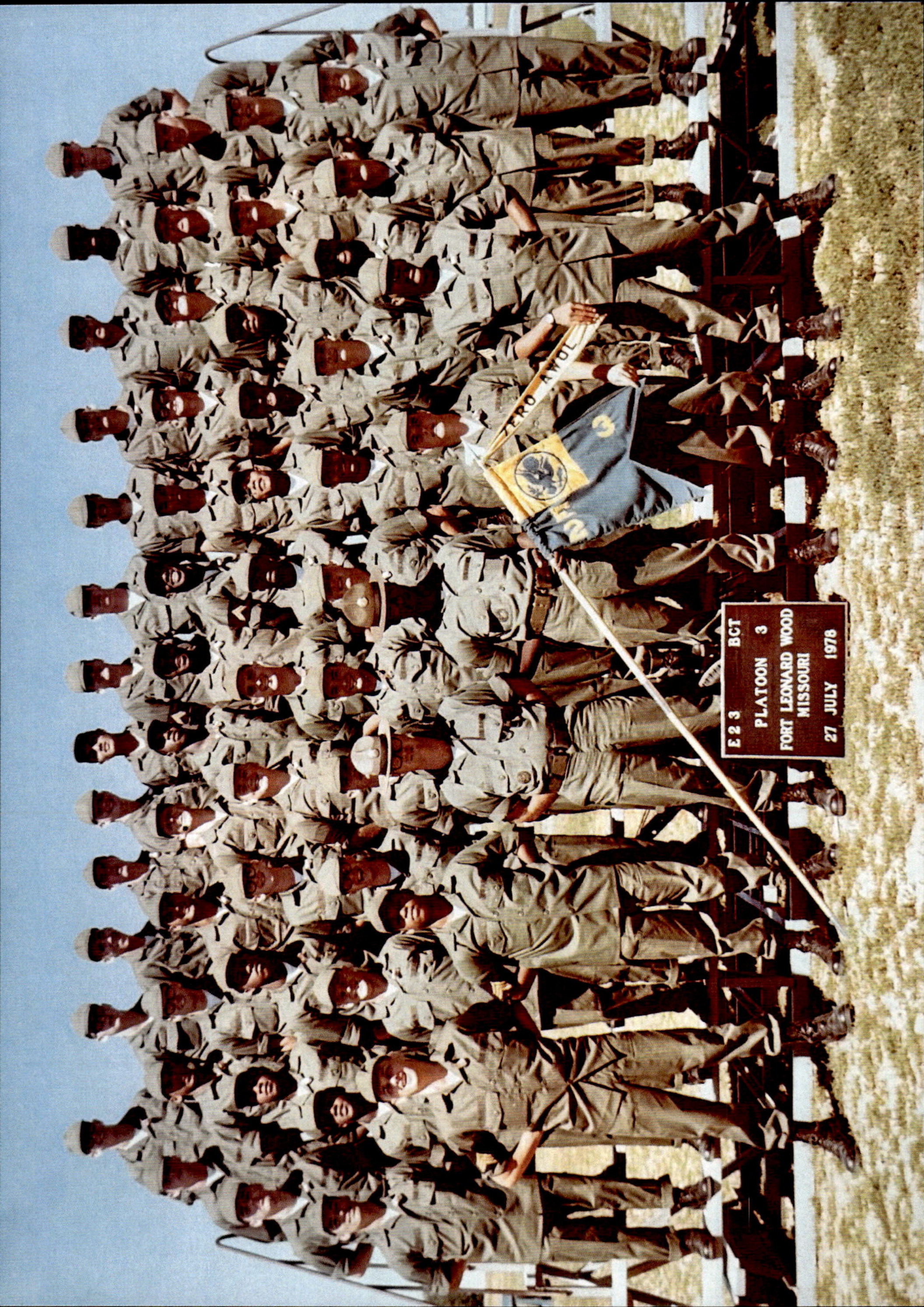
E 2 3 BCT
PLATOON 3
FORT LEONARD WOOD
MISSOURI
27 JULY 1978

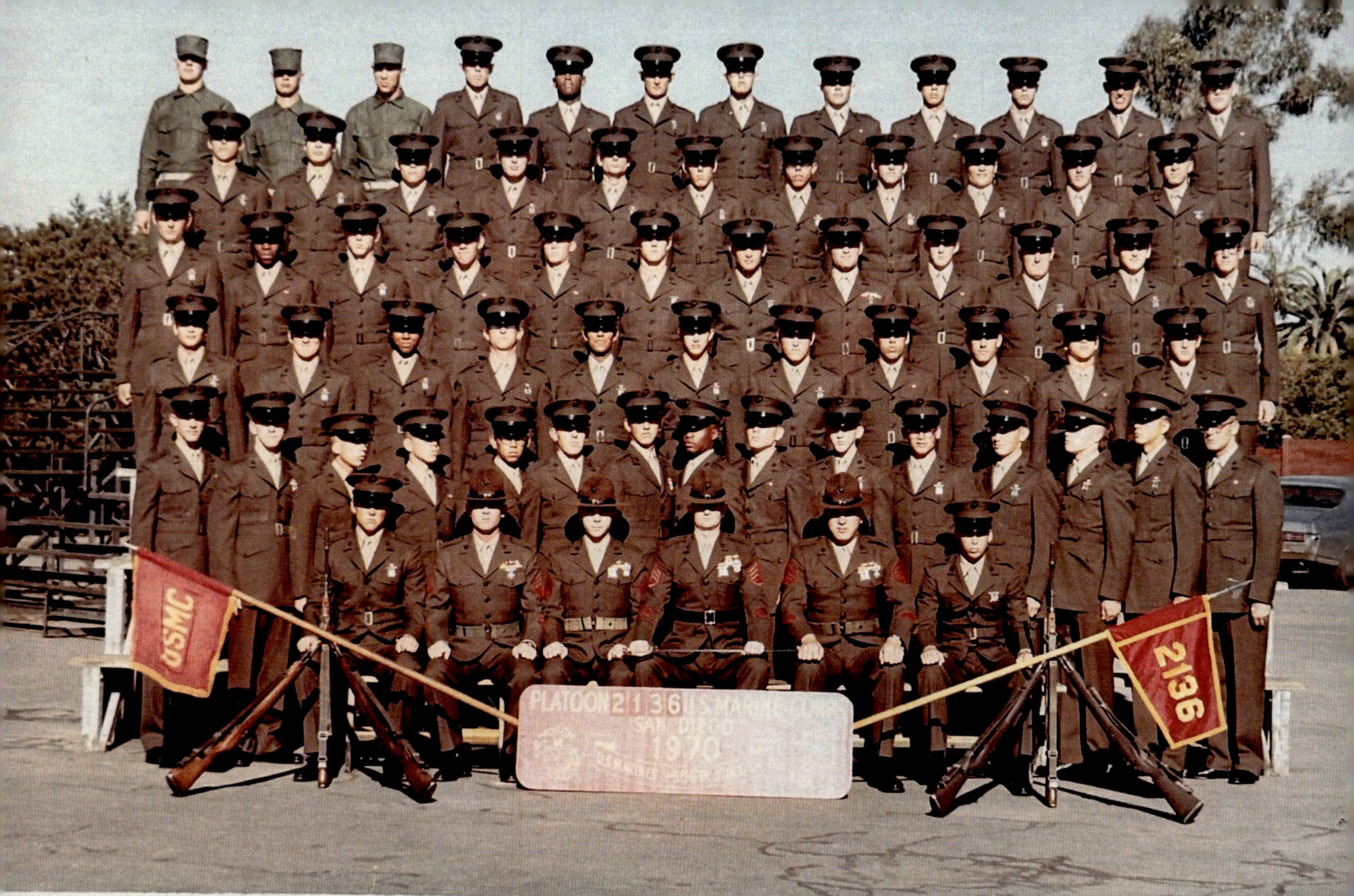

SGT. R. C. COATS SSGT. J. W. SWADER SSGT. D. A. LAW SSGT. C. H. ILNICKY

GRADUATED 7 DECEMBER 1970

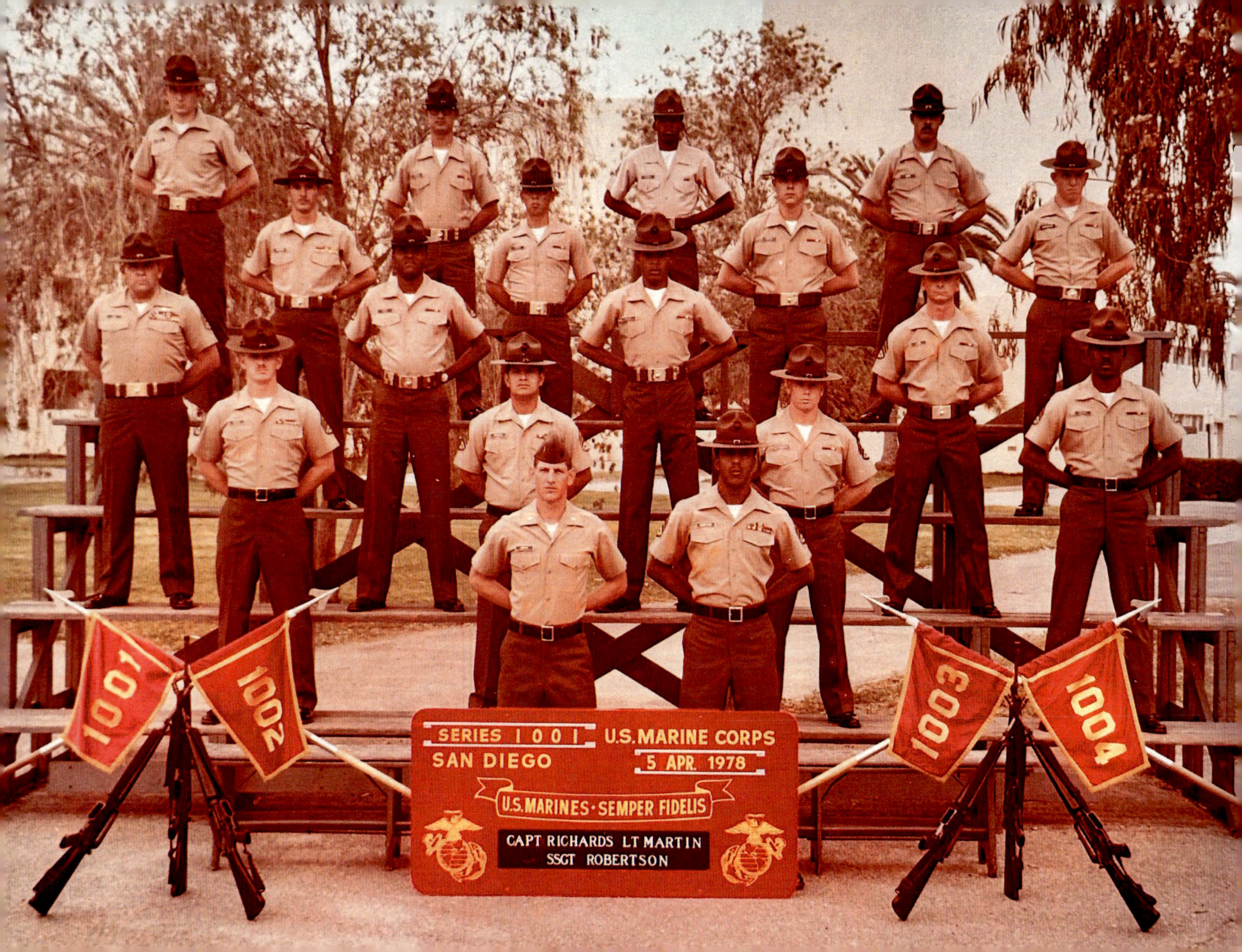
1001
1002
SERIES 1001
U.S. MARINE CORPS
SAN DIEGO
5 APR. 1978
U.S. MARINES · SEMPER FIDELIS
CAPT RICHARDS LT MARTIN
SSGT ROBERTSON
1003
1004

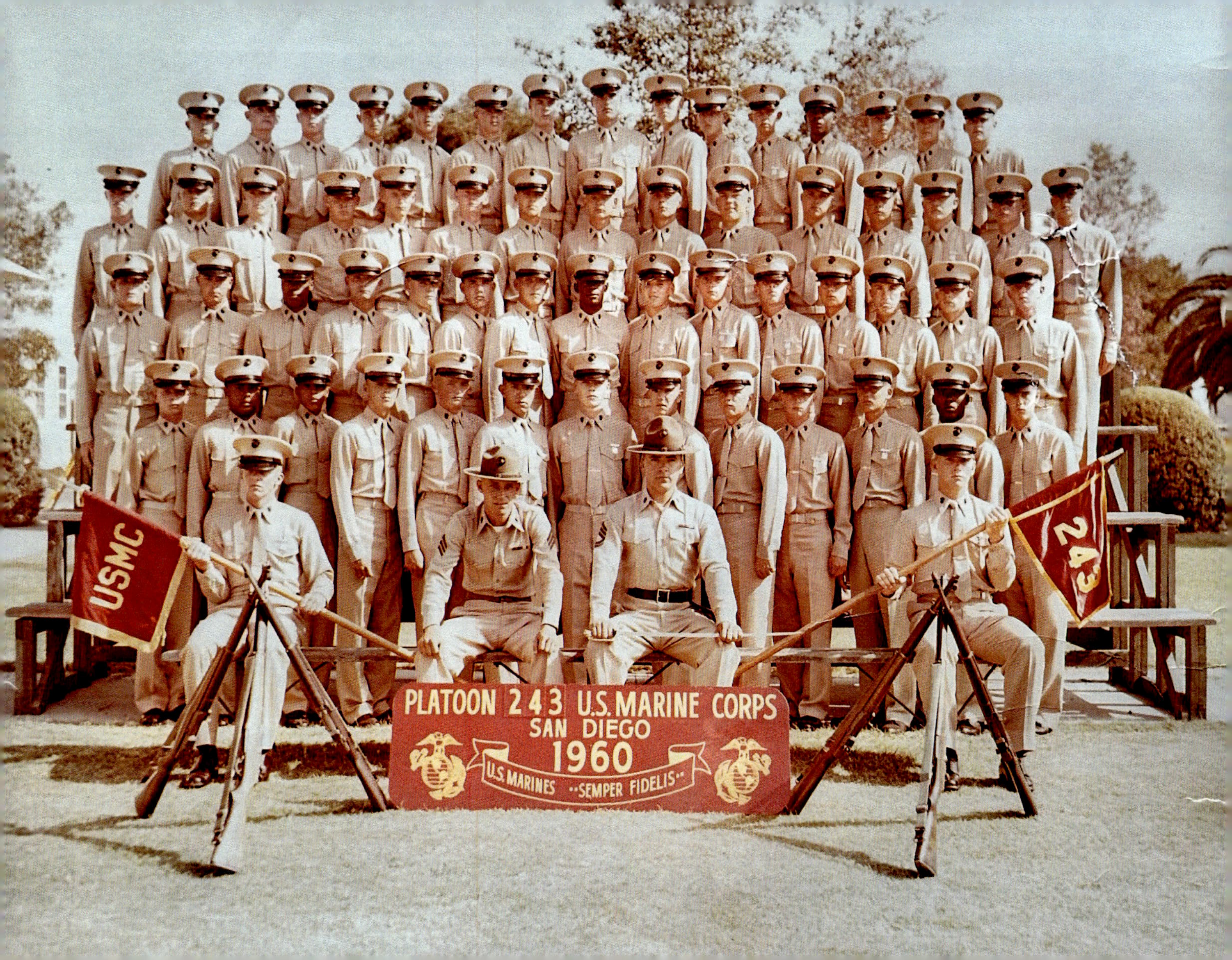
USMC
243
PLATOON 243 U.S. MARINE CORPS
SAN DIEGO
1960
U.S. MARINES ••SEMPER FIDELIS••

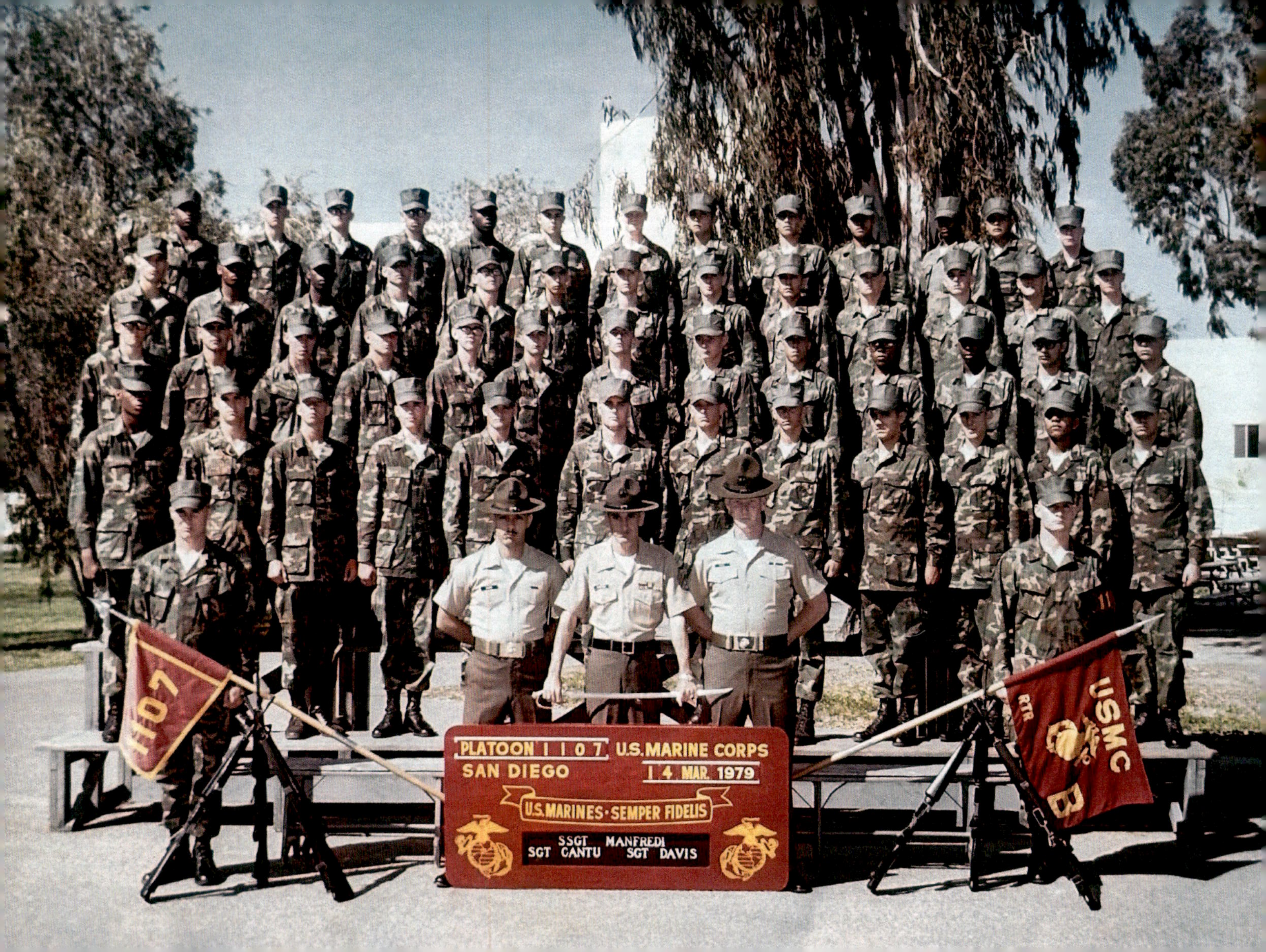
PLATOON 1107
U.S. MARINE CORPS
SAN DIEGO
14 MAR. 1979
U.S. MARINES · SEMPER FIDELIS
SSGT MANFREDI
SGT CANTU SGT DAVIS
1107
USMC
B
RTR

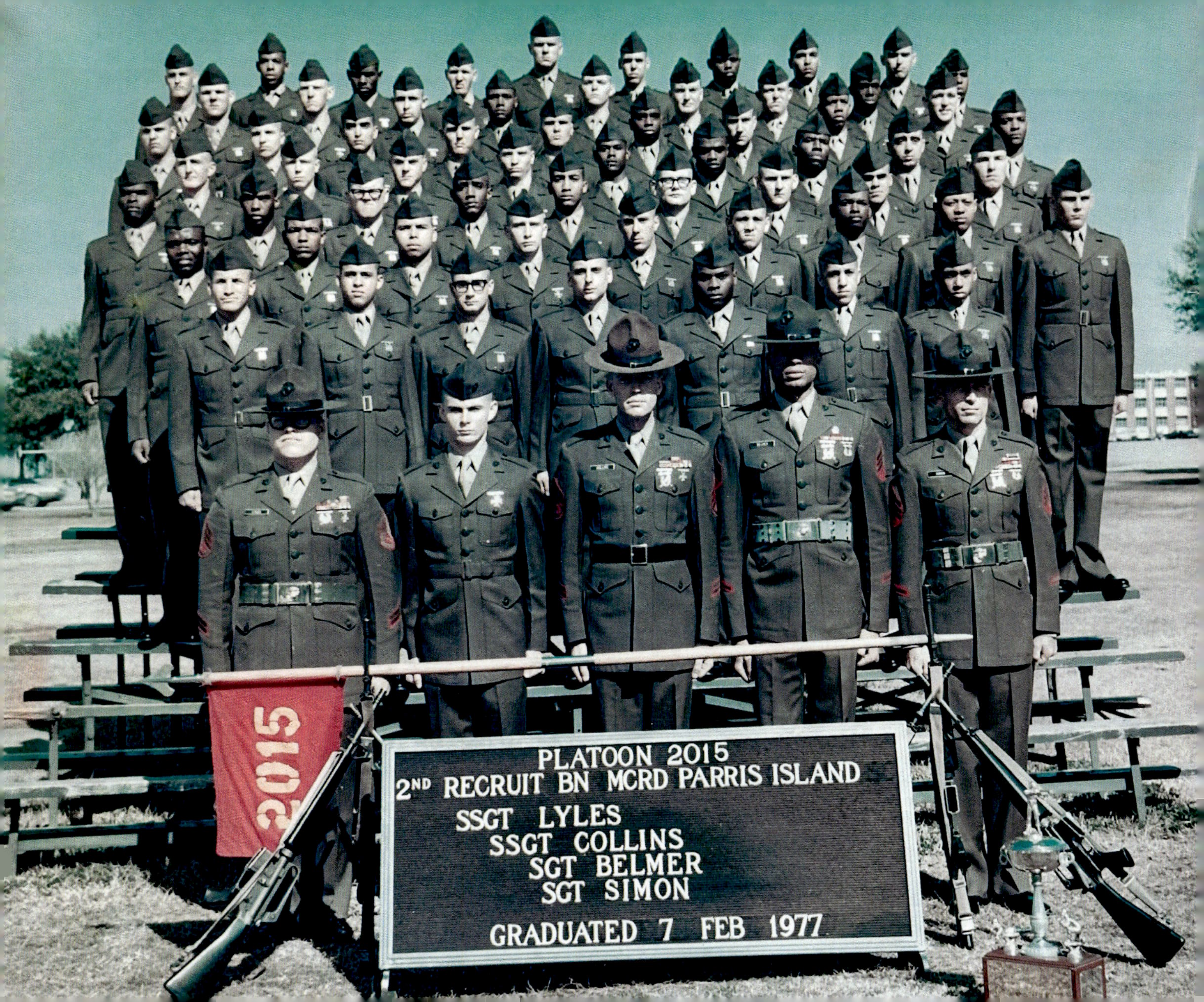
2015
PLATOON 2015
2ND RECRUIT BN MCRD PARRIS ISLAND
SSGT LYLES
SSGT COLLINS
SGT BELMER
SGT SIMON
GRADUATED 7 FEB 1977

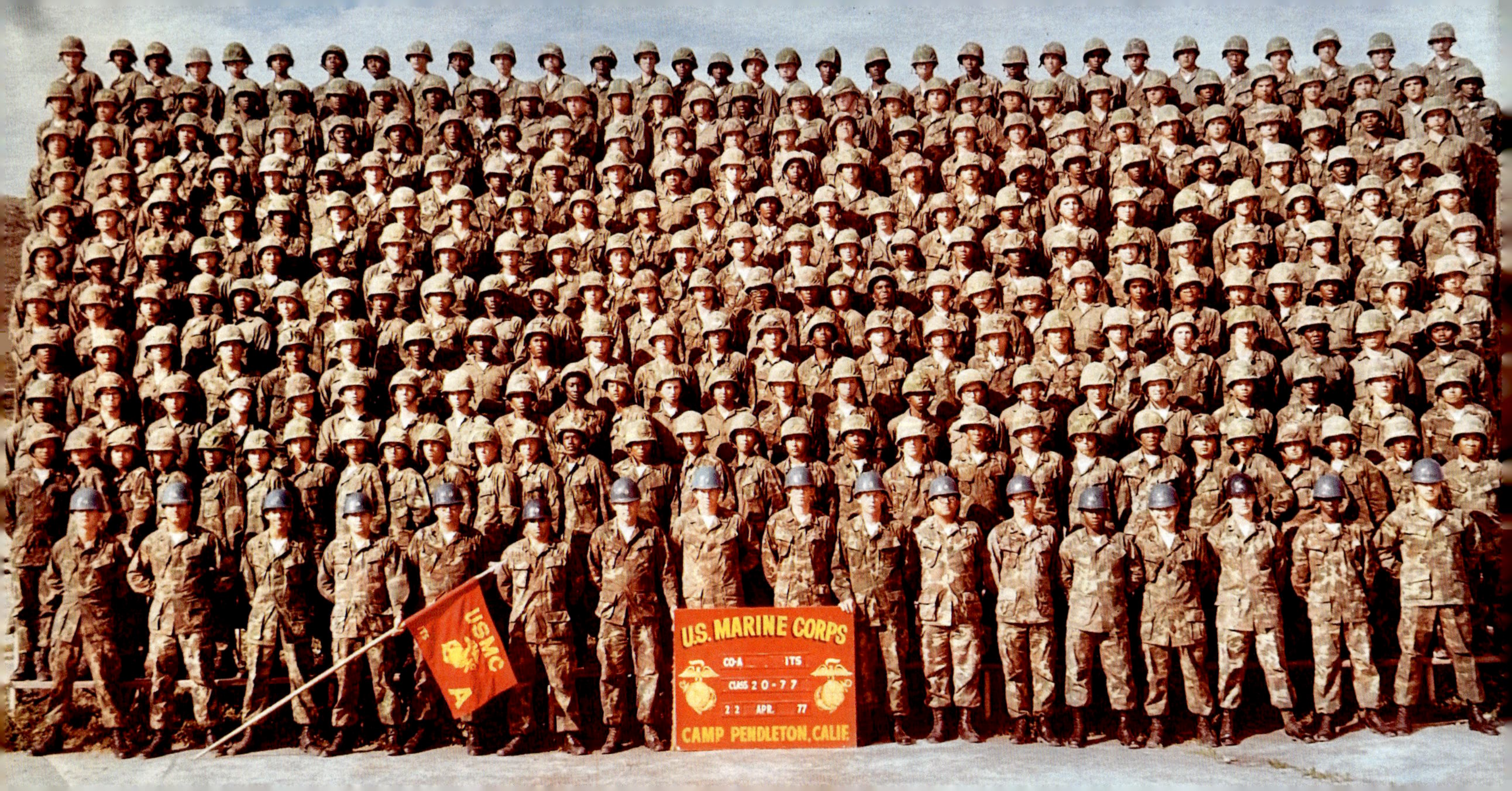
USMC
A
U.S. MARINE CORPS
CO A
ITS
CLASS 2 0 - 7 7
2 2 APR. 77
CAMP PENDLETON, CALIF.

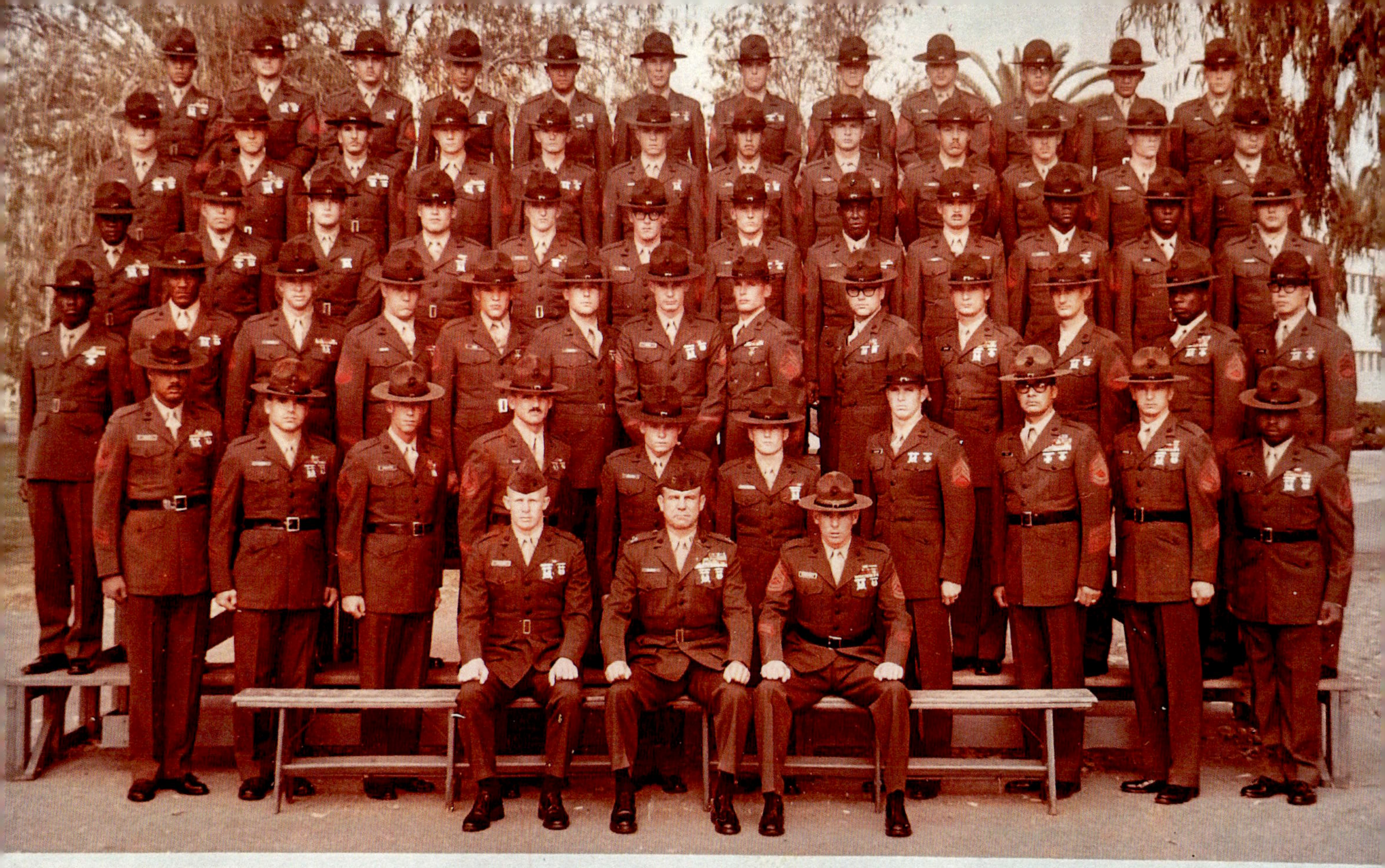

D.I. SCHOOL CLASS 1-78 MCRD SAN DIEGO, CALIF.

LT. M.B. KESSLER, ASST. DIRECTOR, CAPTAIN H.J.WERNER, DIRECTOR, GYSGT. H.G. OVERSTREET, CHIEF INSTRUCTOR

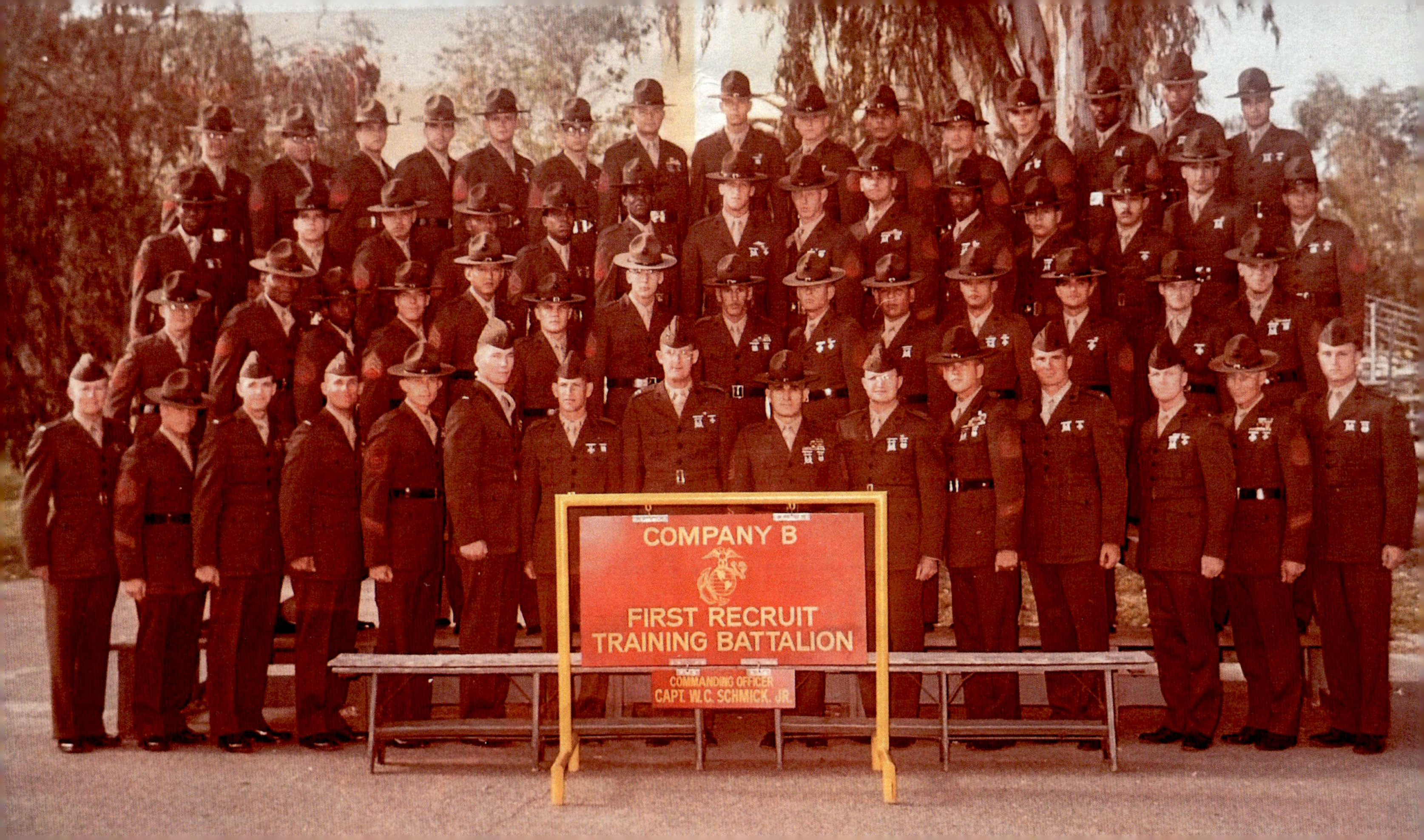

COMPANY B
FIRST RECRUIT
TRAINING BATTALION
COMMANDING OFFICER
CAPT. W.C. SCHMICK, JR.

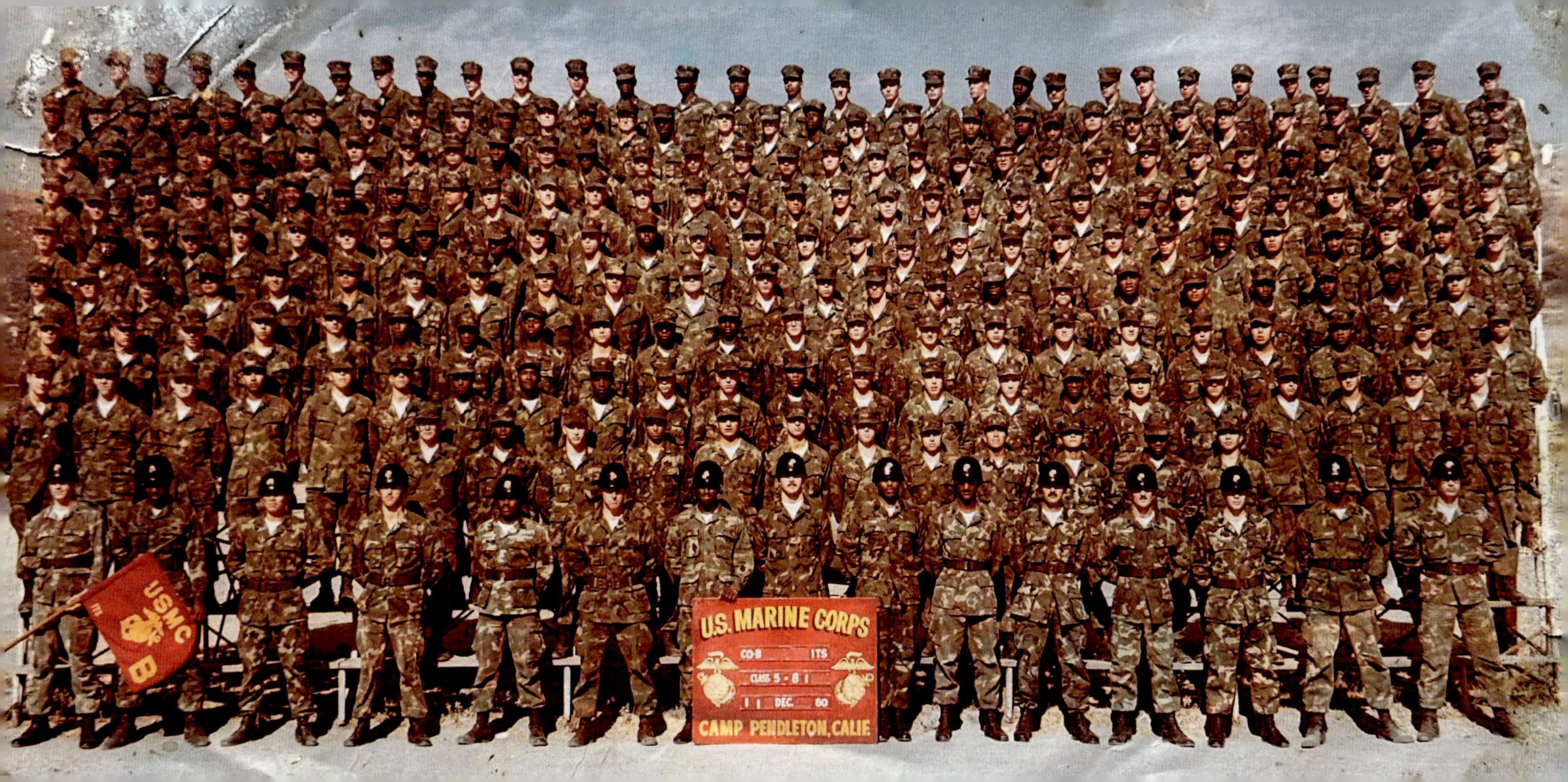
U.S. MARINE CORPS
CO-B ITS
CLASS 5 - 8 1
1 1 DEC. 80
CAMP PENDLETON, CALIF
USMC
B

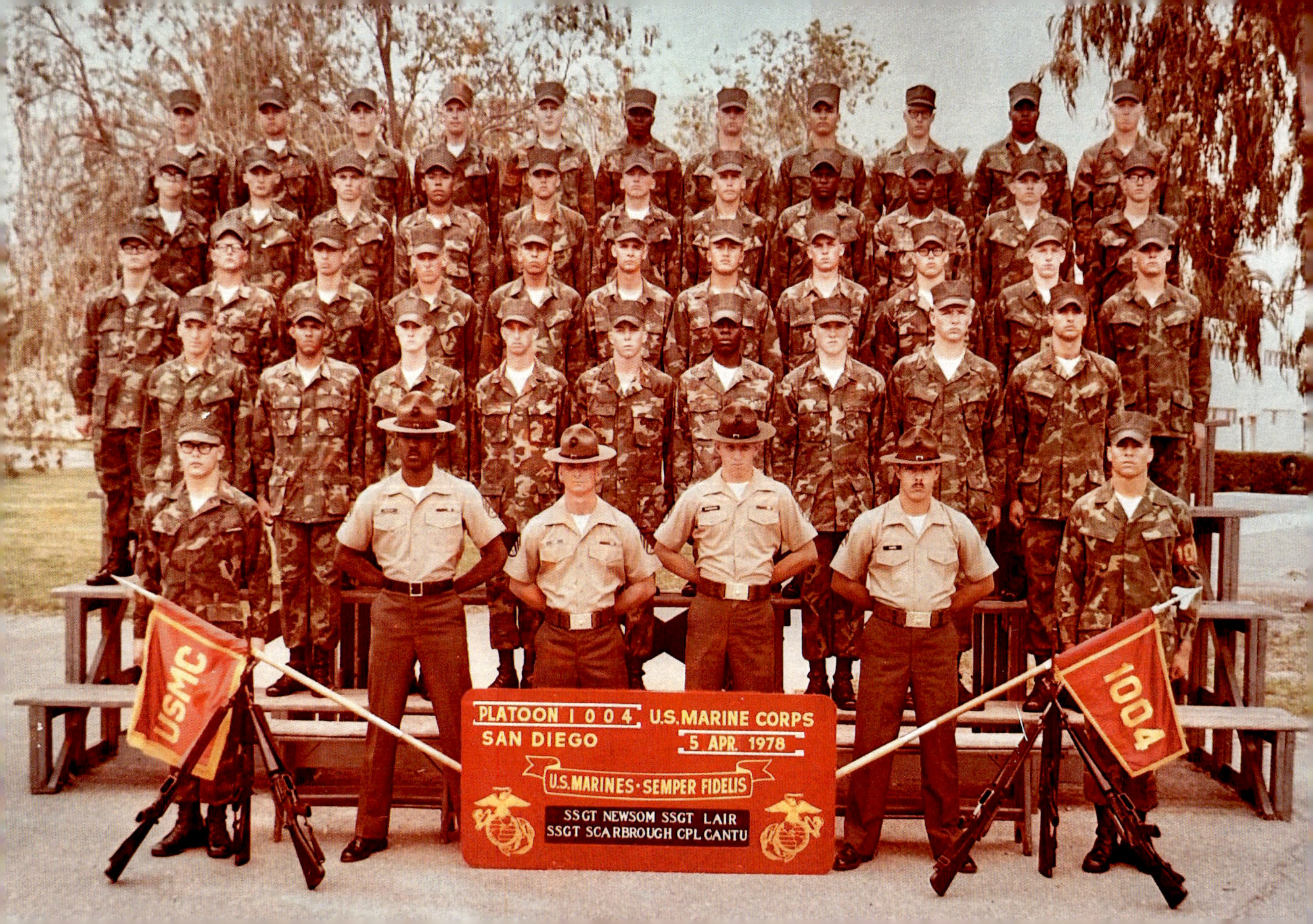
USMC
PLATOON 1 0 0 4
U.S. MARINE CORPS
SAN DIEGO
5 APR. 1978
U.S. MARINES · SEMPER FIDELIS
SSGT NEWSOM SSGT LAIR
SSGT SCARBROUGH CPL CANTU
1004

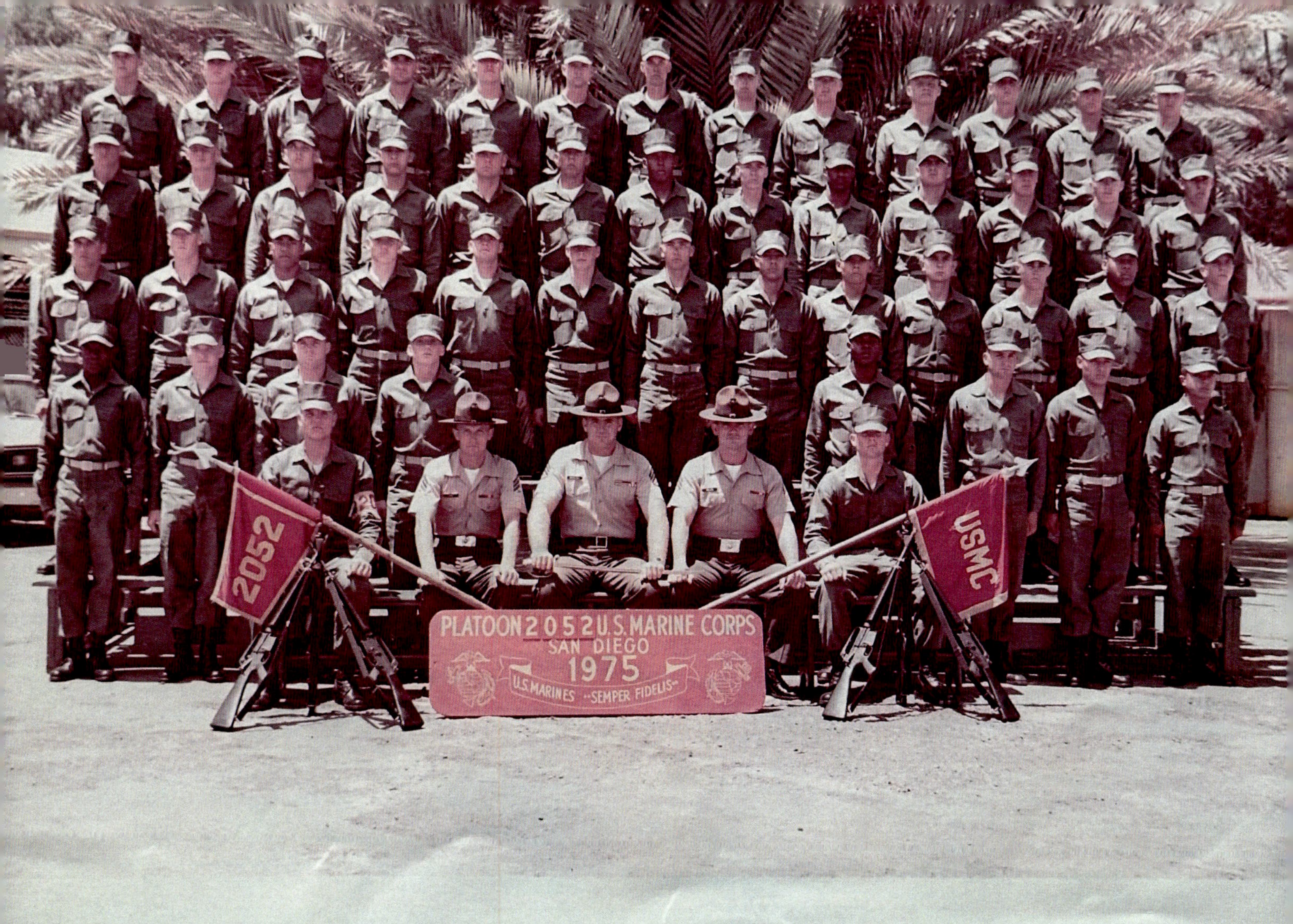

SGT. D. C. LONG SSGT. D. G. ANDREWS SGT. M. T. MULLEN

GRADUATED 1 AUGUST 1975

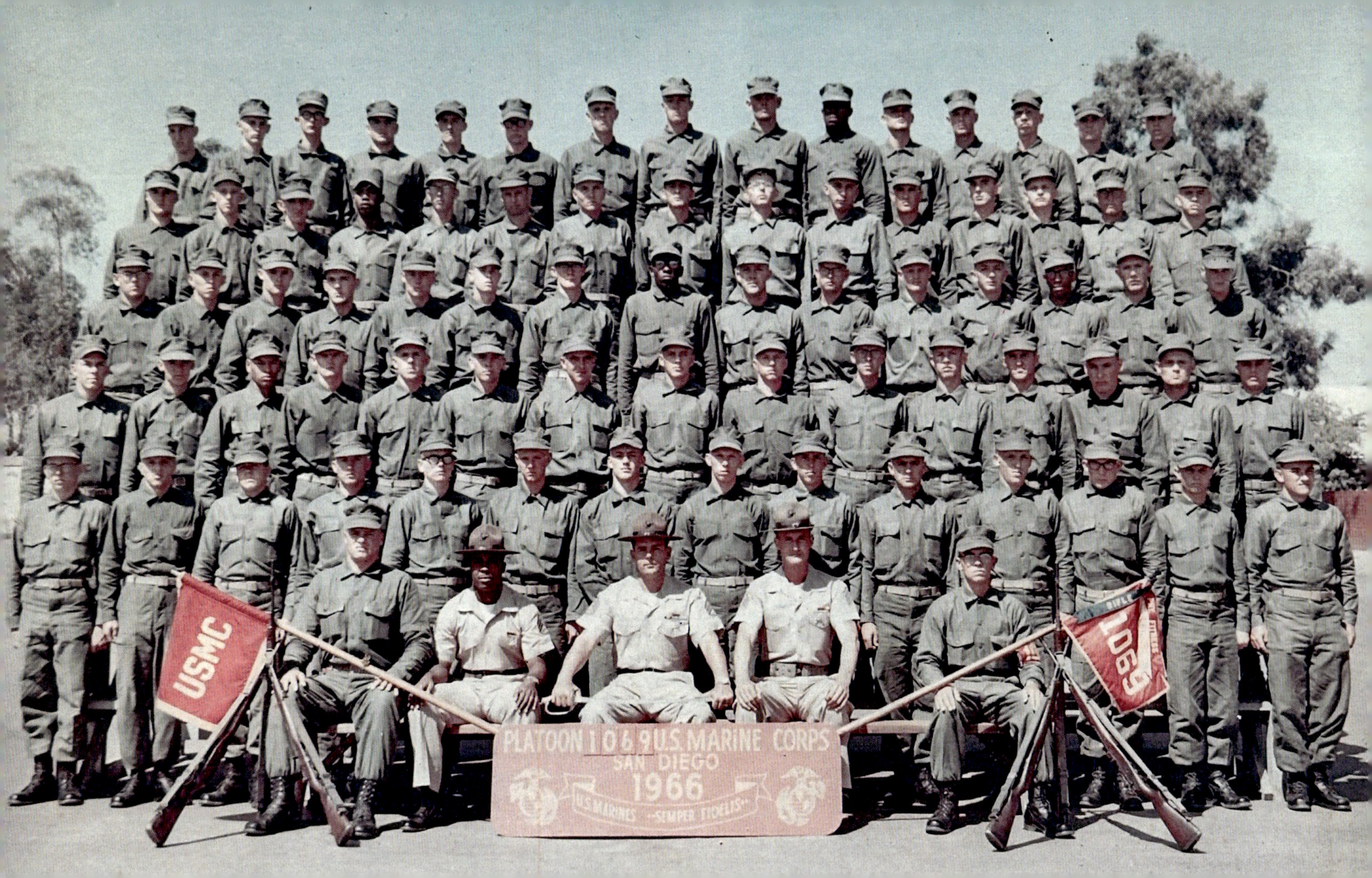
USMC
1069
PLATOON 1069 U.S. MARINE CORPS
SAN DIEGO
1966
U.S. MARINES -SEMPER FIDELIS-

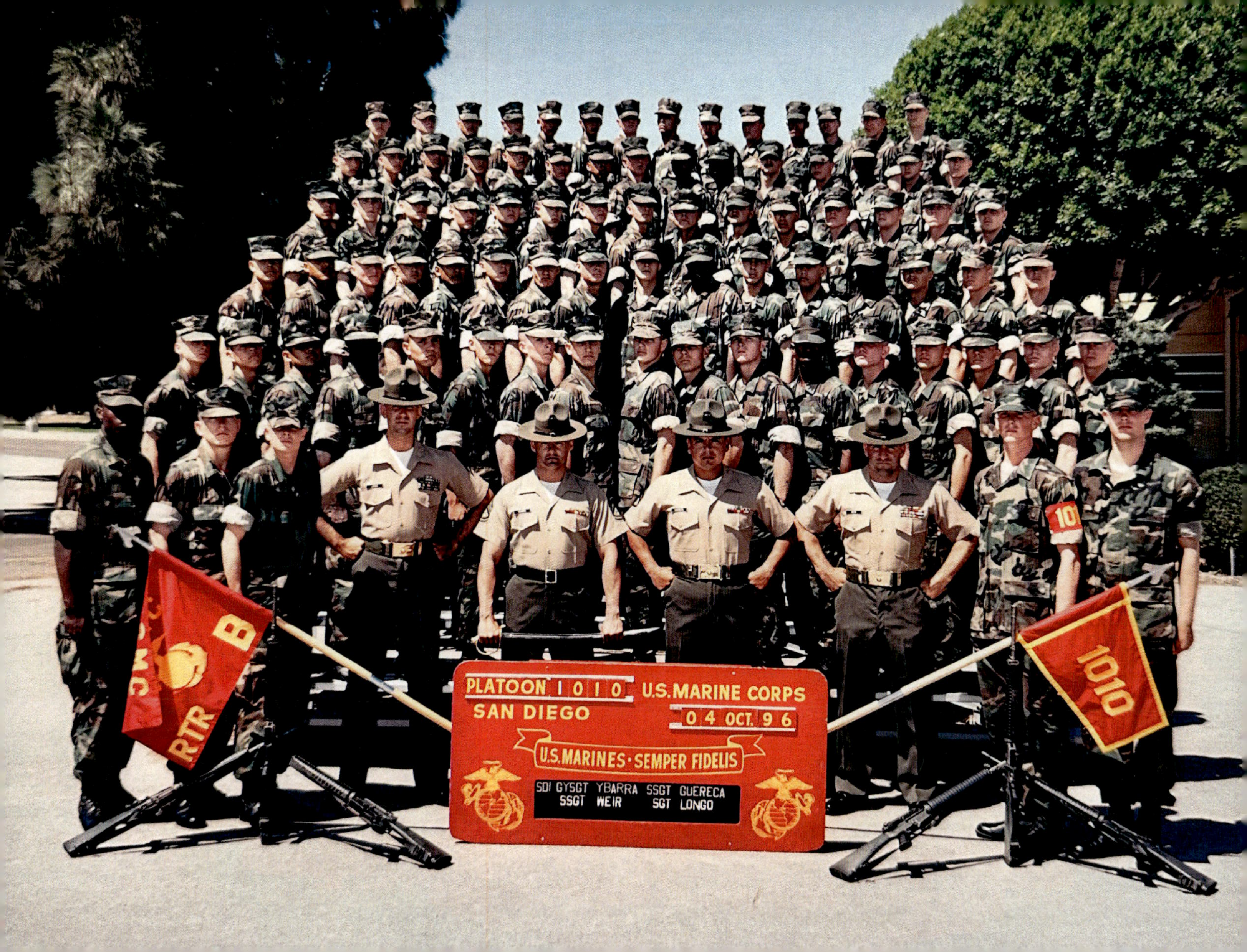
PLATOON 1 0 1 0
U.S. MARINE CORPS
SAN DIEGO
0 4 OCT. 9 6
U.S. MARINES · SEMPER FIDELIS
SDI GYSGT YBARRA SSGT GUERECA
SSGT WEIR SGT LONGO
RTR
B
1010

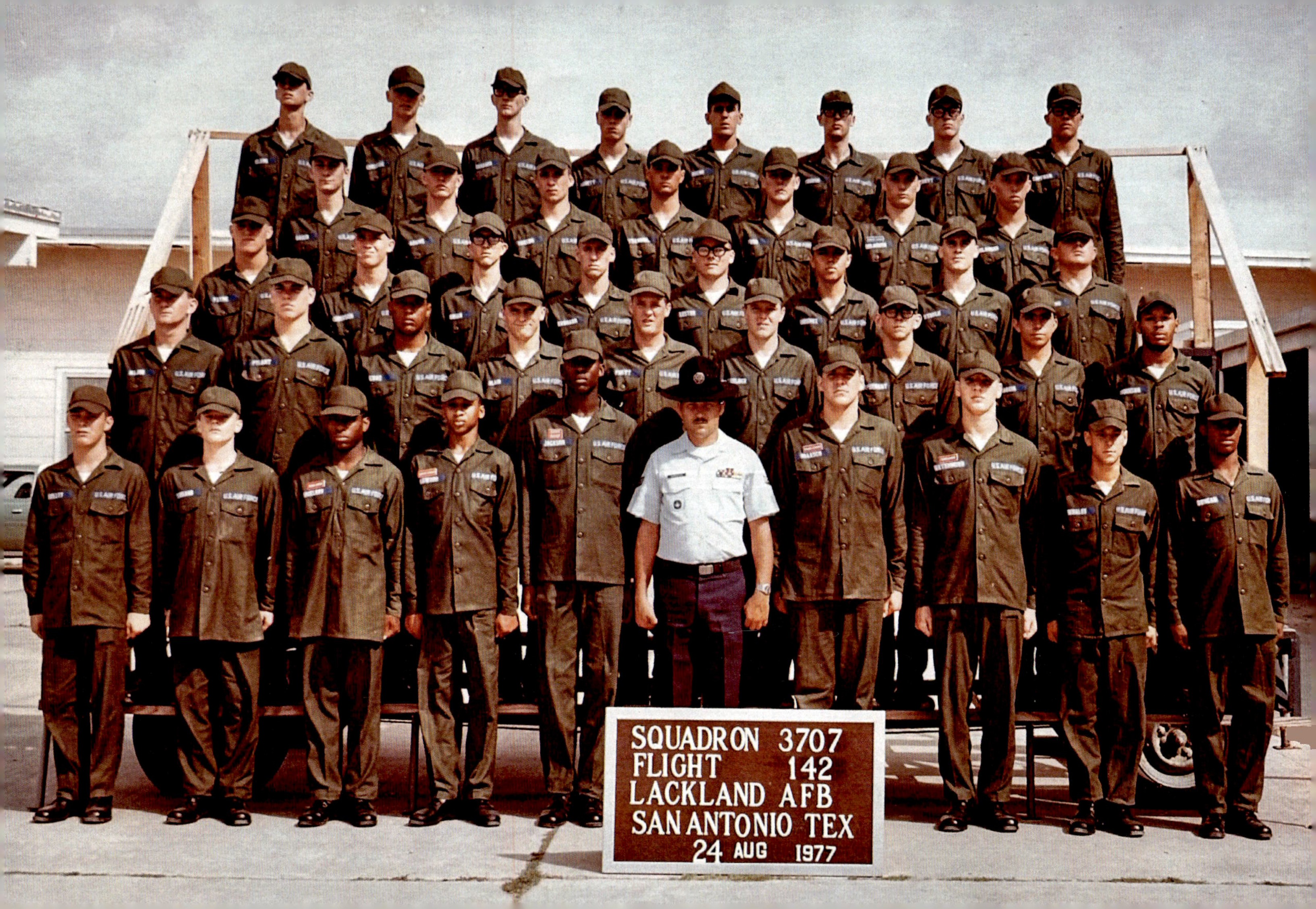
SQUADRON 3707
FLIGHT 142
LACKLAND AFB
SAN ANTONIO TEX
24 AUG 1977

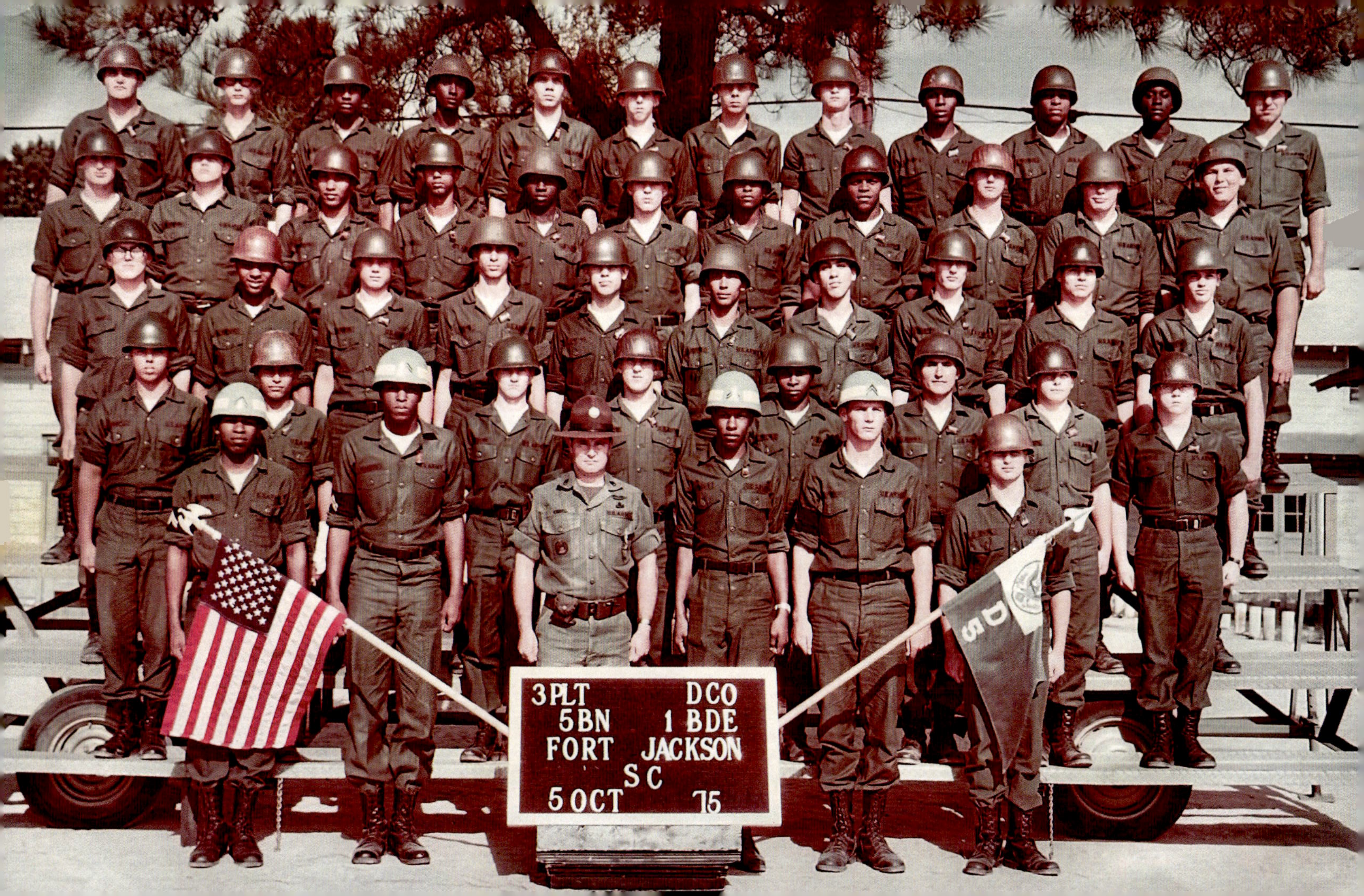
3PLT DCO
5BN 1 BDE
FORT JACKSON
S C
5 OCT 75
D 5

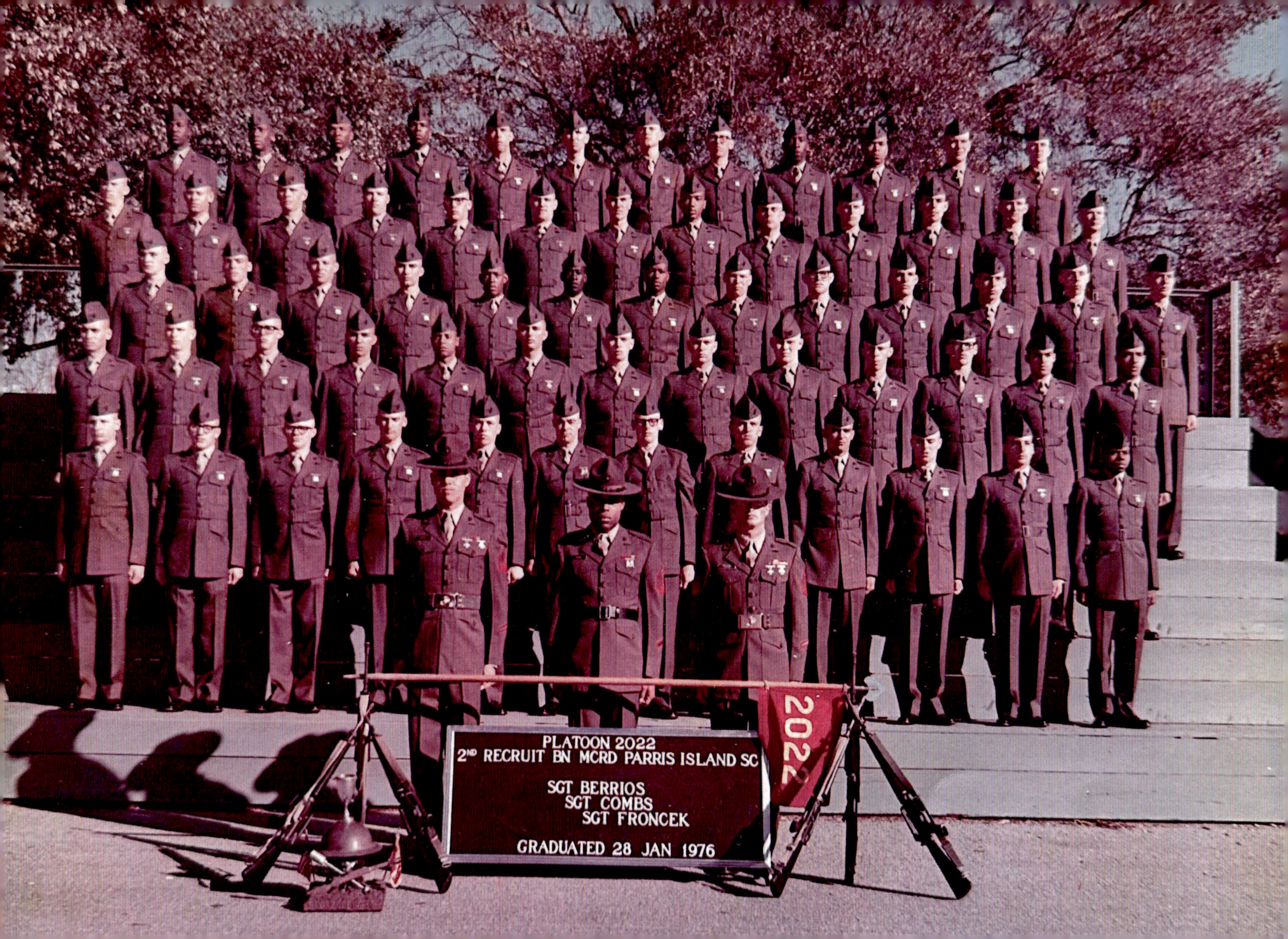
PLATOON 2022
2ND RECRUIT BN MCRD PARRIS ISLAND SC
SGT BERRIOS
SGT COMBS
SGT FRONCEK
GRADUATED 28 JAN 1976
2022

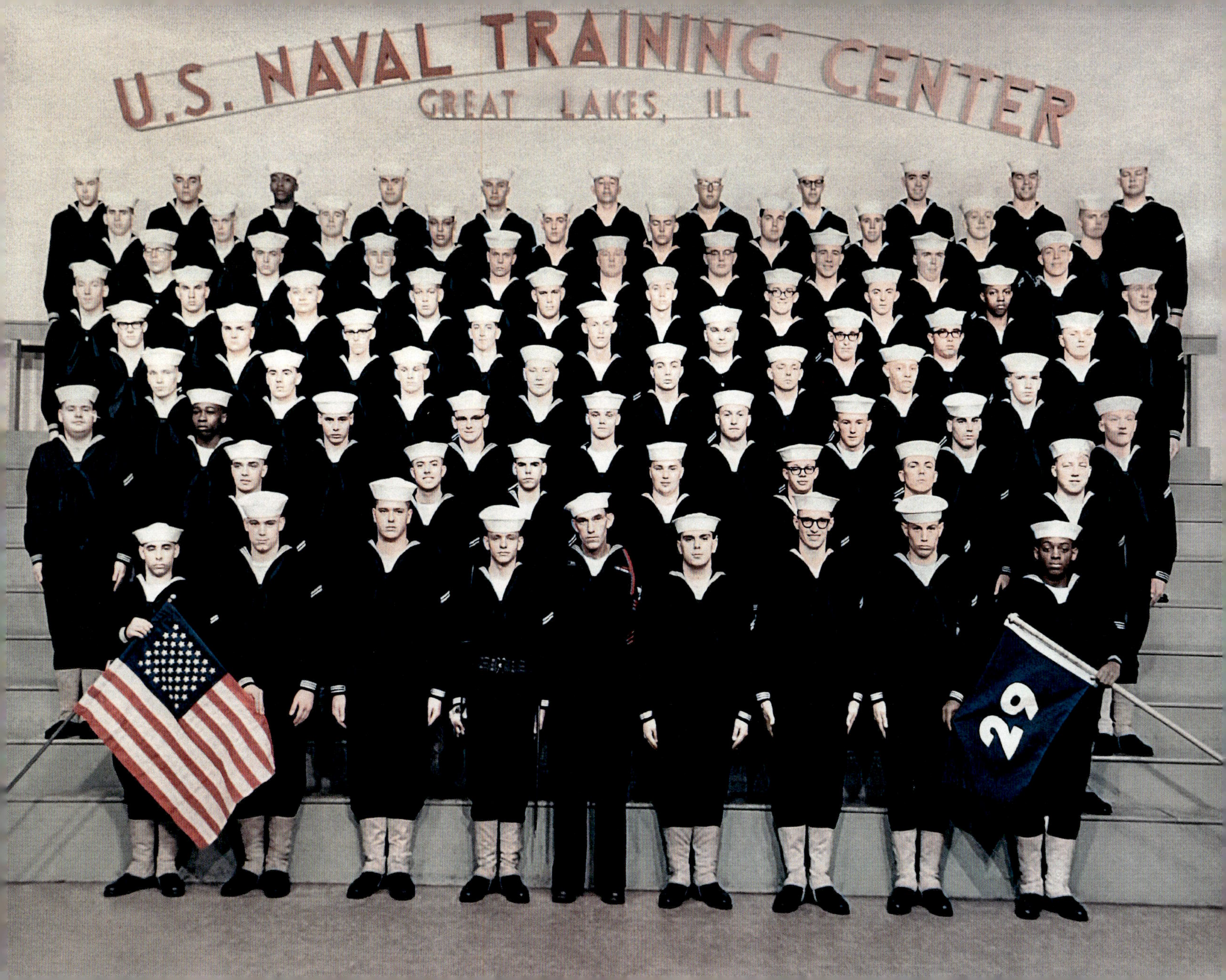
U.S. NAVAL TRAINING CENTER
GREAT LAKES, ILL.
29

About the Author

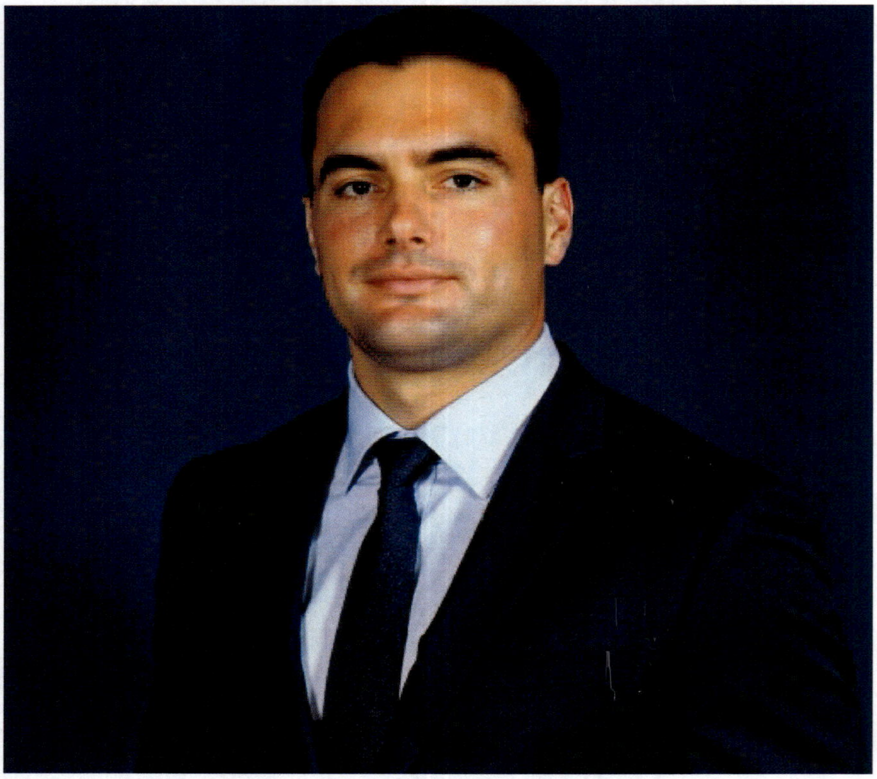

Eric has emerged as a highly accomplished investor and Day Trader, engaging in trading activities with a substantial value of stocks annually. He enjoys the status of a dedicated husband and father, being blessed with four wonderful daughters who hold a special place in his heart. Eric's existence revolves around upholding the virtues of a devout Christian, a devoted husband, and a caring father guided by the grace of God. His journey has been transformative, considering his challenging upbringing, and he now stands as an illustrious testament to the boundless possibilities achievable through faith in God.

Eric joined a criminal street gang as a teenager, and he was eventually arrested and put in a cage at the Long Beach City Jail. He dropped out of Jordan High School in Long Beach, California, in the 10th grade. Eric eventually got on the right path by the grace of God and obtained his GED the same month that he would have graduated. He joined the United States Coast Guard at 18, and after four years of faithful service, he was honorably discharged.

The following is an excerpt from his autobiography: "California Prison Confessions." "It was a typical beautiful summer night in Long Beach, California, and the three young teens fighting in the musty garage starkly contrasted the night's serenity. A teen called Bones was thrown against a wall, and a muscular teenager nicknamed E Double (me) rushed towards him and violently punched him in his lower back. Another youth named Tim Loc held Bones against the wall during this attack. E Double's knuckles landed with a thick-sounding thud as Bones grimaced and moaned in pain from each successive blow.

There's a look of pain and fear in Bone's eyes as E Double mercilessly lands blow after blow to his kidneys. E Double has spent years hitting the heavy bag in his garage, so his punches are devastating for his age. The beating lasted a minute, and the two teens stopped attacking the unfortunate victim. Amazingly, they all shook hands and embraced each other. We were all friends, and this was our modest gang initiation into our small clique called 'Six Feet Under.'"

Our tiny neighborhood gang in North Long Beach, California, had several members of various races: White, Black, and Hispanic. The definition of a criminal street gang according to the California Penal Code is as follows: "As used in this chapter, 'criminal street gang' means any ongoing organization, association, or group of three or more persons, whether formal or informal, having as one of its primary activities the commission of one or more of the criminal acts…having a common name or common identifying sign or symbol, and whose members individually or collectively engage in or have engaged in a pattern of criminal gang activity," California Penal Code 186.22(f).

Some of our associates committed serious crimes such as burglaries, drug sales, stealing cars, carrying guns, vandalism, and assaulting people. Although we were nothing but young punks, we were a criminal street gang according to the California Penal Code.

The FBI did a study entitled "Street-Gang Mentality: A Mosaic of Remorseless Violence and Relentless Loyalty." I want to quote a relevant part concerning young males who do not have a father or other positive male role models in their lives while growing up: "All gang members lacked male role models in their households, and when they lived with their mothers, they were rarely home because of work schedules. Gang members often lived temporarily with various people. All but one gang member reported that one or more immediate family members had a criminal history or abused drugs. Gangs attracted them because of their loyalty, stability, and sense of group identity."

There is an epidemic in this country of boys masquerading as men who don't take their responsibilities as fathers seriously. Many fathers aren't involved in their children's lives; if they are, they are a horrible example of what a real man should be. One of the main factors in young men ending up in the penal system is the lack of a father or, at the very least, a positive male role model when they were growing up. Absentee fathers are causing society to pay a high price for their selfish and callous actions.

Every day, I strive to be the best dad to the four beautiful daughters God has blessed me with. I live by the motto, "Be the man you needed as a child." I will not allow the actions of physically and emotionally abusive people to define who I am or hinder my advancement in life. You can choose to be bitter or better. I strive for excellence because living well is the best response to abusive people.

Eric Sturgess

God bless you and yours.

Romans 10:9-13

For those seeking to connect with Eric or gather additional insights about his upcoming projects, please visit his official website: EricSturgess.com.

Other Books

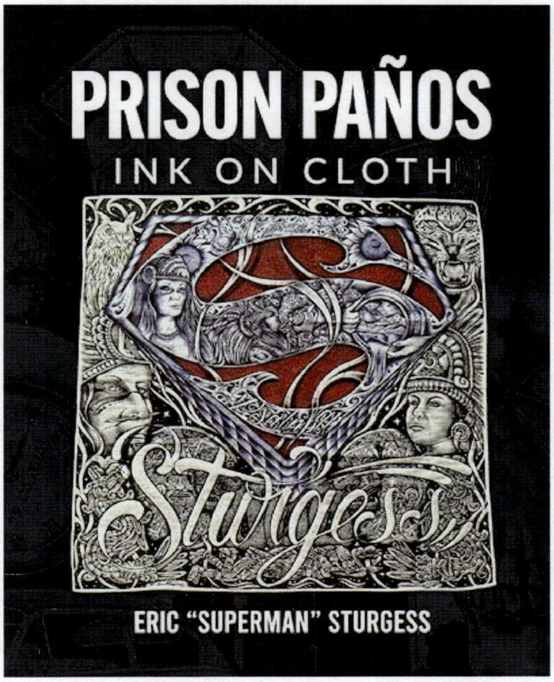

Prison Paños: Ink On Cloth

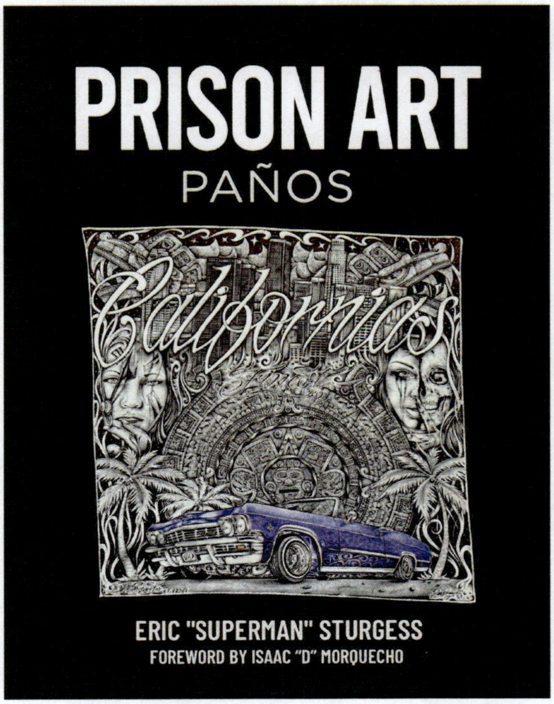

Prison Art Paños

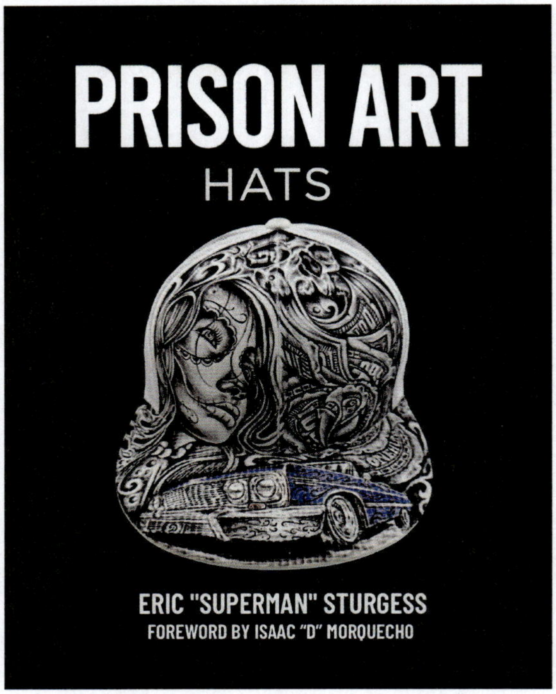

Prison Art Hats

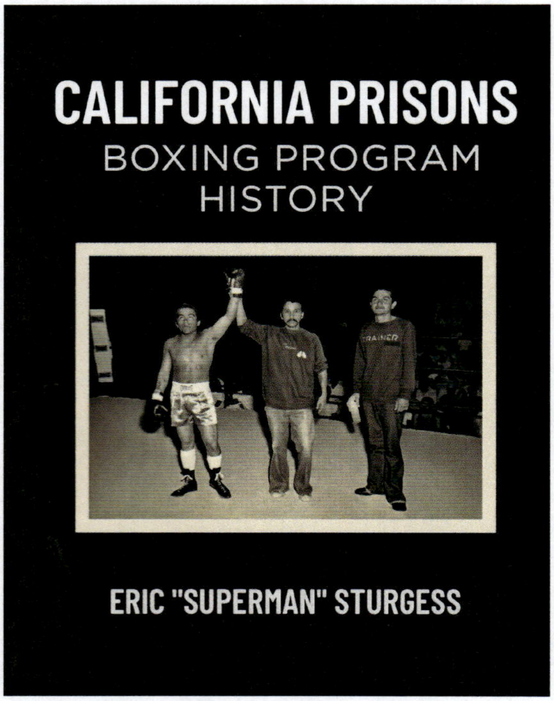

California Prisons:
Boxing Program History

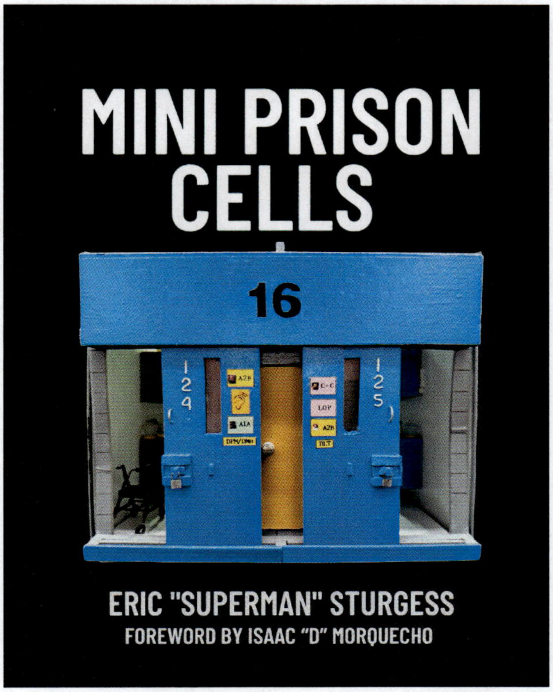

Mini Prison Cells

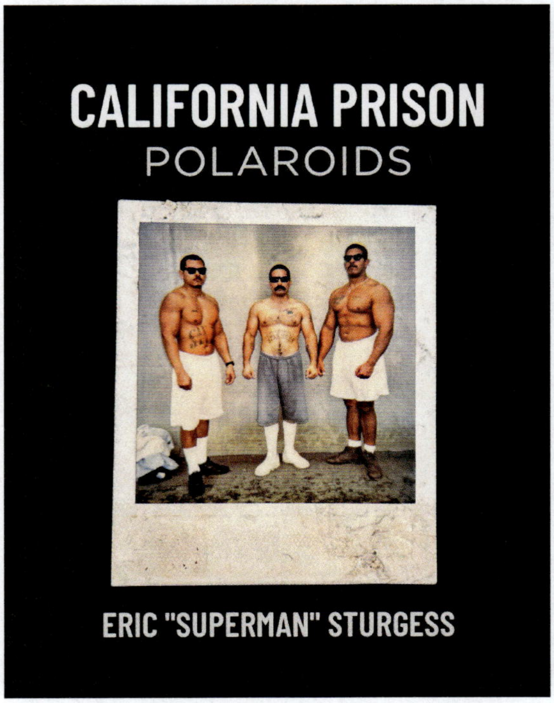

California Prison Polaroids

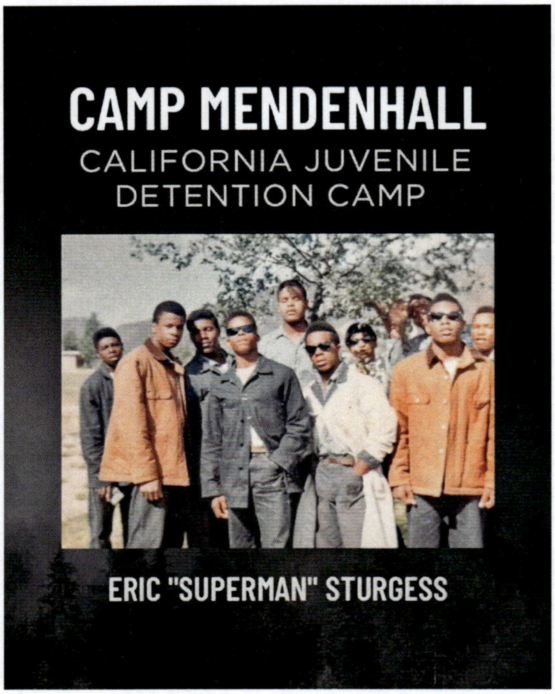

Camp Mendenhall: California Juvenile Detention Camp

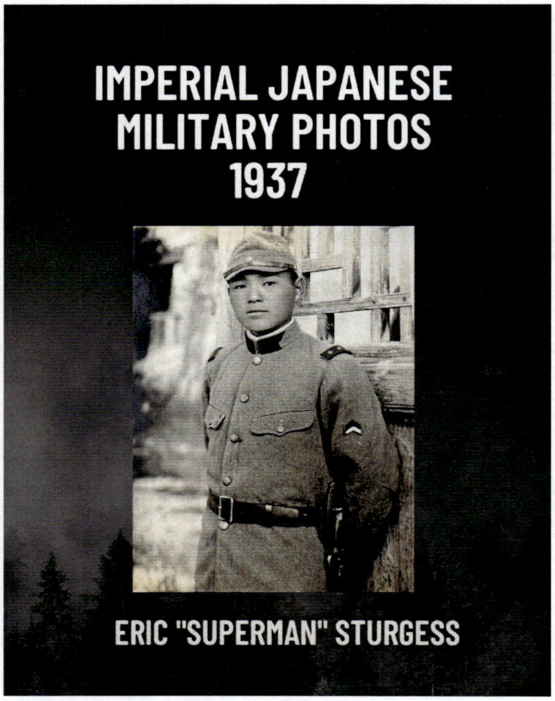

Imperial Japanese Military Photos 1937

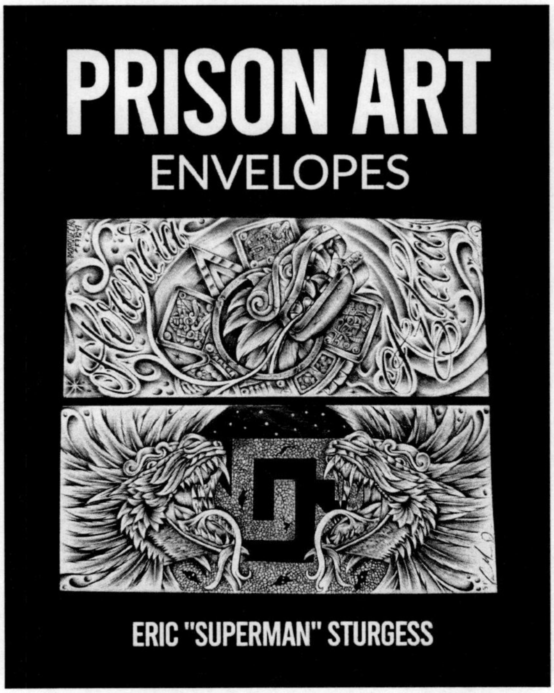

Prison Art Envelopes

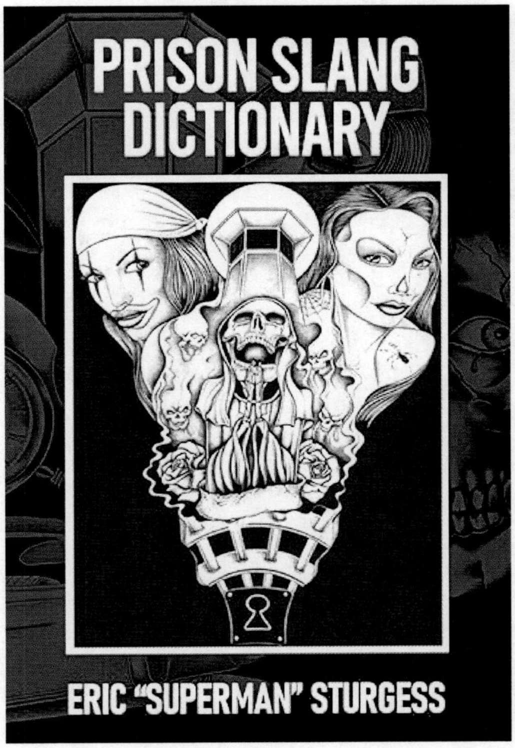

Prison Slang Dictionary

Made in United States
North Haven, CT
05 November 2026

21765358R00024